INNOVATION

TO
FRED (RUTH)

Innovation

A Battleplan for the 1990s

John W. Carson

Gower

Published by
Gower Publishing Company Limited
Gower House
Croft Road
Aldershot
Hants GU11 3HR
England

Gower Publishing Company
Old Post Road
Brookfield
Vermont 05036
USA

British Library Cataloguing in Publication Data
Carson, J.W.
   Innovation: a battleplan for the 1990s
   1. Industries. Innovation
   I. Title
   338'.06

ISBN 0 566 02785 2

Printed in Great Britain by Woolnough Bookbinding Limited, Irthlingborough, Northamptonshire.

6-3-91 ?

CONTENTS

Contents

FOREWORD

Karl von Clausewitz revolutionized the practices of warfare in the 1800s; his lasting memorial was a handbook on this subject. John Carson has taken a military theme as his central metaphor for innovation, and has also provided us with a handbook – a plan of campaign for the industrial and commercial battlefields of the 1990s. He offers new weaponry (techniques), tactics ('how to do it' hints), and strategic principles (theoretical guidelines). This foreword offers a situation report on the wider terrain of innovation studies up to the present time.

Early skirmishes

Before 1950 innovation was widely confused with invention. When mentioned, the processes were assumed by economists to be the expeditionary forces of growth. Thus the generals of these forces were inventors-cum-entrepreneurs, that is, the heroes of industrial revolution.

In the 1950s attention shifted from the generals to the weaponry. Technology became the new hero which swept aside resistance and effectively won the battles against the resisting forces. But observers of innovation practices reported successes from a quite different approach, which went with the war cry 'market pull'. The 1960s were to see furious encounters between technology push advocates, and their

market pull adversaries. Eventually an uneasy truce emerged, uniting the parties in what became known as 'contingency theory', in which the line of command for any innovation encounter had to be alert to specific circumstances (that is, contingencies) such as counter-thrusts from the enemy, surprise technological breakthroughs, secure outposts or niches, or dynamic regrouping opportunities.

Innovation in the 1970s and 1980s

In the 1970s there were interesting campaigns which established a new fusion of theoretical ideas and practical sorties. In Europe Jan Buijs (now Professor of Innovation at the University of Delft) directed a most successful one. From his trained force of process consultants he allocated one to each of a series of small firms. The introduction of structured techniques and special innovation task forces led to substantial reported innovation gains.

At about the same time John Carson was pioneering the use of SCIMITAR, his first ideas machine, in industrial organizations. Again it was demonstrated that a structured approach could transform a company and permit substantial gains in market territory.

Each of these approaches was founded in the principles of action research – learning by doing. Each confirmed the significant impact of a trained adviser, the innovation corps, and a need for drills and routines backed up with imagination and flexibility. Indirectly, the exercises reinforced the predominant contingency mood of the times.

Prospects for the 1990s

It is a brave man who predicts future battlegrounds and ultimate victories. The reported successes of John Carson's methods have made startling reading. However, as an academic I was initially sceptical about the claims for his first innovation system, SCIMITAR. Nevertheless my doubts were banished by the confirmed results, in terms of new product triumphs in market places for organizations following his approach.

As someone interested in the theory of innovation I welcome the fascinating new methods and results reported in this book. They will go some way to strengthening links between theory and practice. Practice has always outrun theory: not until many years after David defeated Goliath were the principles of centrifugal force established. We still do not know exactly how and why firms overlook 'obvious' opportunities; nor do we know the scope and limitations of the innovation techniques. It is less than clear which mechanisms trigger creative thought.

I look forward to the 1990s in the expectation of progress in these areas.

Tudor Rickards
Center for Studies in Creativity
State University of Buffalo
New York
Manchester Business School
England
1988

Preface

The Managing Director was dismayed by our analysis of his company's future:

The rate of change in both technology and marketing in your sector has accelerated rapidly in the last five years. A new product range launched today can expect a profitable life of only four years.

87% of your last year's profits came from two product ranges launched between eight and ten years ago. Both have already started to decline and neither will be profitable two years from now.

The new product range you attempted to launch two years ago has failed because you misunderstood the market, fell short on technical performance and were beaten by a full year by competitive solutions.

Your Managers are not ideas people. Collectively they could only name eleven ideas for significant new business lines – the success rate in your industry is one launchable winner from every fifty ideas.

Our analysis of your Business Development Meetings showed that between 90 and 95% of all that was said was mere regurgitation of known facts, no new thinking was evident and no new conclusions were reached. (The rate of progress toward a conclusion in a discussion is inversely proportional to the number of people involved – you involve 17 in your Development Meetings.)

Few of your Managers seemed to have had any training in innovation techniques and the atmosphere in your Company seems to militate against creative thought. Cost effective innovation requires a different group of management skills to the management of existing business and a different psychological approach.

Your Managers spend their days arguing and competing against one another, trying to maintain their positions and the status quo of their departments. The result is that their time and effort is being frittered away on urgent trivia while no time is being found to view longer term issues of vital importance to your Company.

Our conclusion is that unless both management styles and Company ethos are changed rapidly, you will fail to manage the major changes now apparent in your industry and go out of business inside the next five years.

Admittedly, this is an extreme example. However, we preface this book in this way to pose the question: 'How well are you managing the future of your organization?'

ix

Acknowledgements

To the 12200 practising managers who have taken part in SWORD programmes for industrial and commercial new business generation.

To the 800 companies in the UK, Western Europe, North America and Asia who have been involved in SPEAR, SABRE, SCIMITAR, BLADE, RAPIER or CUTLASS programmes for real growth.

To the science park movement world-wide, through contact with which I have had a unique opportunity of studying entrepreneurship.

To business schools in the UK, Western Europe and North America from whose workers in the field of innovation I have learnt my trade.

To Tudor Rickards, my mentor in creativity and its applications in the management of innovation.

Finally, to Diane and Ruth for the typing and graphics and secretarial input to this book.

My thanks to you all.

J. W. C.

Synopsis

1 *What is this book about?*
This book is about generating real growth for business and industrial companies. As such, it contains management systems which generate significant profit improvement. These systems are geared to the management of change in these fast-moving times, which is more difficult than managing the status quo. To survive in the 1990s it will no longer be sufficient to run a stable company; all companies need to change continually to keep pace with their business surroundings. This process of change must involve innovation.

2 *Who should read it?*
The primary readership targets of this book are practising business and industrial managers, directors and executives. However, it should also be of interest to those who study business and industry. So, if you are either a practising manager or a student of business, this book should be relevant to you. It contains not only theoretical bases but also practical systems and case studies on innovation.

3 *The title*
This book is called *Innovation, A Battle Plan for the 1990s* because it provides a battleplan for innovation. 'Innovation' is the one-word description for the complex process of change and advancement in business and industry. The subtitle places the book in terms of the development of management systems for innovation. It is in fact a complete rewrite and update of *Industrial New Product Development – a Manual for the 1980s,* and as such presents new systems for innovation developed in the last decade.

4 *The format*
This book is set out in the form of a manual in the hope that it will be used: business people rarely read books! The manual is divided into parts, each covering a specific task in innovation. Within each part chapters cover problems, practical systems, mental approaches and proven programmes. As such, the practising manager or student of business can thumb through the manual and find relevant weapons for specific tasks. We have kept the chapters short, 1000–2000 words each, so that the busy manager can read them quickly.

5 *Why the weaponry acronyms?*
We believe that to manage innovation successfully you must cut through a great deal of resistance. The management of change, by definition, involves some disruption of the status quo and it is human nature to resist this. We therefore believe that managers need specific weapons in their armoury, and we have sought to provide these in this book. Each part presents a specific weapon for this armoury, and is subtitled with the appropriate acronym as an aide-mémoire.

6 *The basis of the book*
This book is based on the systems devised, tested and applied by a

specialist consultancy in business and industrial innovation – Sword Innovation Services Limited (hence the armoury!). Over the last decade Sword has worked successfully with just over 1000 business and industrial client companies. The systems used and the output produced, upon which detailed statistics have been compiled, constitute the database of this book.

As such, the systems presented have been used successfully by over 12000 practising managers, working in over 200 different business and industrial sectors, on a worldwide basis. To date our statistics show that these systems have created over 500000 relevant new business concepts for clients and, from these, more than 10000 ventures have been captured.

7   *Is it theory or practice?*
This is a practical book, but it presents both theory and practice. In each part we outline the difficulties observed in conventional approaches, then provide the theoretical background and follow this by presenting specific manage-ment systems for the application of that theory. Finally, operational pointers and case studies are presented to back up the practice. However, it should be stated from the outset that many of the theories used cannot be proven outright. Many draw on current business and industrial psychology theories which are still in the experimental stage. Nevertheless, we present them, for your consideration, on the basis that they have stood the test of time over the past decade in practical application.

8   *How to use the book*
We hope that this book is sufficiently digestible to be read from cover to cover. More important, its part and chapter structure can be read as a reference manual to provide specific systems on demand. We have tried to present the systems in sufficient detail for readers to use them. However, because new management systems require both learning and application, this book can provide only a starting point – practice, as always, is vital.

PART I

INTRODUCTION TO SYSTEMS FOR INNOVATION

SYSTEMATIC

   WEAPONS for

      ORGANIZATIONAL and

         RESOURCE

           DEVELOPMENT

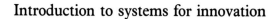

Part 1 of the manual examines the need for change in business and industry, provides a theoretical basis for the study of the changing dimensions of companies, and sets the scene for specific approaches to systems for the effective management of innovation.

1

The dimensions of business and industrial change

Business and industrial change

In this chapter we examine the phenomenon of business and industrial change and provide a framework for its examination. In particular, we address such topics as the need for change, the psychology of change and the difficulties which change creates in business and industry. The chapter is divided into sections dealing with definitions, problems, approaches and systems.

Starting point

Before examining systems for innovation in business and industry, we should first define what we mean by business and industry. The business and industrial spheres are highly complex and much work has been done to provide frameworks for their understanding. The starting point for all the management systems we present is the simple concept that business and industry can be examined in a multi-dimensional framework.

The dimensions of business and industry

Our studies, through consultancy with business firms and industrial companies,

have led us to believe that business and industry are six-dimensional systems. By this we mean that specific businesses or industrial companies can be defined using six dimensions.

- Source
- Process
- Market
- Time
- Finance
- People

We regard each of these as separate dimensions. They interact together to produce remarkably complex and problematical conjunctions, but they can be examined separately. Most of our innovation systems use three-dimensional models, in which the three most appropriate dimensions are drawn from the six listed above, to suit a specific situation.

By examining opportunities for changes within these three-dimensional models, we are able to provide a systematic approach for the various stages in the innovation process. In this way, we can model the management of change: see Figure 1.1.

Are people a dimension?

Do people fit into our multi-dimensional approach? The question is whether 'people' constitute a dimension. We have built models in which people do constitute a separate dimension, and equally successful

3

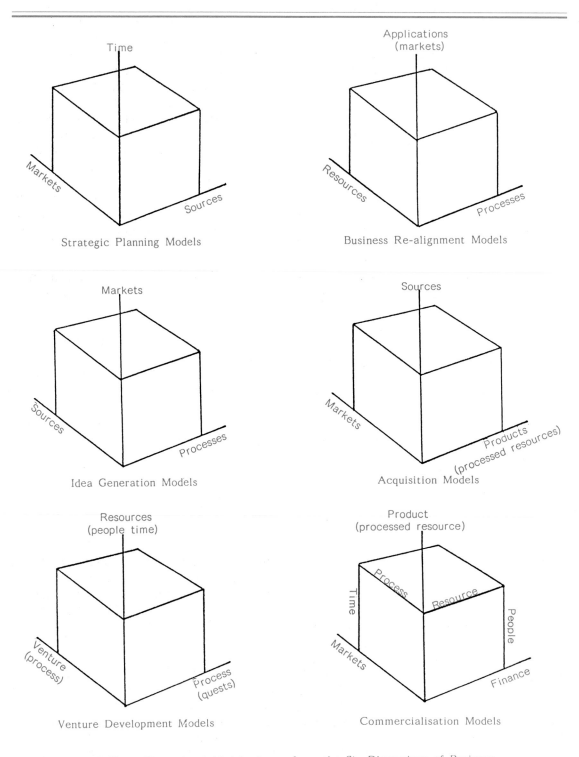

Examples of Three Dimensional Models drawn from the Six Dimensions of Business

Figure 1.1   The dimensions of business and industry

models where their input is implied on each of the other dimensions. We prefer not to be dogmatic. However, we stress the role of people throughout the innovation process and recognize that their psychology is of paramount importance.

Theoretical and practical dimensions

We have built well over 2000 business models for client companies. Each model is specific to the company concerned and to the innovative process being applied. To confer the necessary degree of specificity it is vital to choose the right dimensions. For any one model we choose the most dynamic three dimensions out of the six available. Specific models are defined in each part of this book. As will be seen, to achieve the necessary specificity we often have to use combination and derivative dimensions, rather than the theoretically pure dimensions listed above.

The choice of specific derivative dimensions for the modelling of particular stages in the innovation process will be explained in each part of the book.

So what is business, innovation and the management of change?

Our definition of business is therefore a six-dimensional complexity involving sources, processes, markets, time, finance and people as the dimensions. Not all six may be represented in any specific business or industrial sector. For example, an industrial manufacturing company would certainly have a processing dimension to its operation, whereas a business offering a service might not. Conversely, businesses in the financial sector may indeed have a processing element as they process money – an example of deriving a specific dimensional meaning.

Models can be either static or dynamic. If they involve time they are dynamic, and these are the ones in which we have specialized. It is possible to build a model of a company in a static form and to use it perfectly satisfactorily in the management of the status quo. However, for reasons which will become apparent later, we believe that few companies will survive the 1990s unless they can go beyond the maintenance of the status quo and embrace change.

Therefore we will present specific dynamic modelling approaches which are useful in the management of change in business and industry.

Our definition of business and industrial innovation is simply the successful management of change in these environments. Innovation means doing something new, in a business sense, and – hopefully – doing it cost efficiently. Our initial premise is that to manage anything as complex as business and industry on a cost-efficient basis first requires a framework for understanding the dimensions involved.

More than systems

Once a framework is established for understanding the dimensions involved in the innovation process, we have a starting point. However, we will argue that modelling is a necessary but by no means sufficient approach.

Innovation involves the management of change and it is human nature to resist change. It is vital, therefore, to take psychology into account at all stages in the innovation process. Psychological barriers are sufficiently powerful to defeat the best management systems if presented in isolation.

We believe that business and industrial psychology will be vitally important in the

5

1. Left Brain/Right Brain Hypothesis

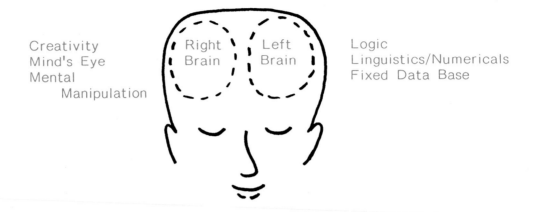

Creativity
Mind's Eye
Mental
    Manipulation

Logic
Linguistics/Numericals
Fixed Data Base

2. Front/Back of Mind Hypothesis

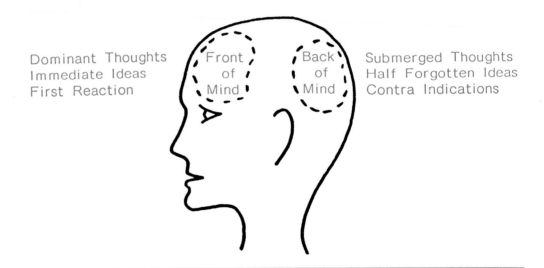

Dominant Thoughts
Immediate Ideas
First Reaction

Submerged Thoughts
Half Forgotten Ideas
Contra Indications

Figure 1.2   Psychological hypotheses used in the innovation process

1990s, but we acknowledge that the current state of understanding provides only hypotheses not laws. Nevertheless, our experience over the past decade has shown that the available hypotheses are effective in breaking down the resistance to change experienced in innovation. The psychological hypotheses we have used are shown in Figure 1.2.

Each part of this book contains one of these simple psychological approaches to the management of change in the specific innovation step being considered. This use of psychology is intended to support the implementation of the systems presented.

We know that these psychological hypotheses are gross oversimplifications. However, we use them in business and industry because they work. We also freely acknowledge that the techniques we use add nothing new to the science of psychology; what we do say is that the techniques are new to most business and industrial companies.

Concluding comments

We have met many managing directors who 'expect' their managers to manage the rapid changes in technology and market places now surrounding their companies, without recognizing that this is a far more difficult task than running a stable, slow-moving business – which is what the managing director has experienced. Very few practising line managers have been trained to comprehend the multi-dimensional task they are expected to perform and even fewer have any insight into the complex psychology of change.

How many 'training courses' have you attended, that focus on the future for your company rather than some aspect of managing current business?

2

Who needs innovaton?

In this chapter we underline the difficulties involved in innovation and then justify the effort required to overcome these difficulties by defining the need for business and industrial change.

Pitfalls

Is it at all surprising that we meet difficulties in the management of change within a six-dimensional complexity? Companies need only get one of the dimensions wrong and the business may change from a stable into an unmanageable situation. Worse than this, as we shall see later, the various dimensions can interact in a most perverse way!

It can also be argued that many of the difficulties and disasters experienced in innovation stem from the people involved rather than the other dimensions. As we said earlier we regard people not just as another dimension but as the vital psychological element. Handled wrongly the people aspect can be more devastating than any other dimension. 'Additudinal problems' in the staff of businesses and industrial companies are a prime pitfall in the path to innovation.

We will examine these pitfalls in Chapter 3, but first let us justify the effort that will be needed to overcome the difficulties. Stated simply, we are asking 'Why bother?'.

Must we innovate?

If the management of innovation is so difficult, can it be avoided? We argue that all companies change; indeed it seems that all systems change over time, whether natural systems within the biosphere or humanistic systems in society. Certainly, all business and industrial systems seem inevitably to change. Moreover, changes within these spheres are far more rapid than either of those in nature or society generally. And at the present time business and industry appear to be changing at an accelerating rate.

Very few industries have remained stable over the last decade. Individual companies within these industries have seen their existing business base eroded ever more quickly. Within many spheres of technology companies have faced the choice of either changing rapidly or going out of business. We cannot believe that competition based on technological and commercial innovation will become less severe in the 1990s: indeed we argue that the need for innovation will become paramount. For the vast majority of businesses and industries change is inevitable, and innovation must be embraced if the company is to survive.

Managing change

We believe that business management can modulate the speed of change. By this we

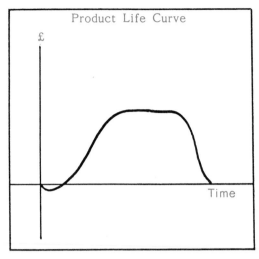

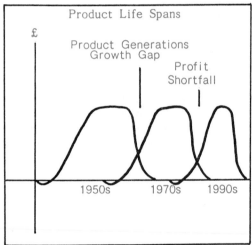

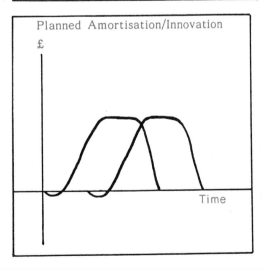

Figure 2.1 Change curves

mean that change can be managed so that it is a smoother process than it would otherwise be.

If management recognize the inevitability of change they should see the need for both specific management systems and psychological approaches that help smooth the process; see Figure 2.1.

If the corporate ethos is backward looking and maintenance of the status quo is the dominant management endeavour, however, change is likely to be anything but smooth. In this unfortunate situation the change curve would be a sequence of plateaux and precipices. The plateau is where management succeed in the short term in maintaining the status quo only to fall from stability down a business precipice. Ultimately, this form of change in business is likely to lead to the death of the specific company and its replacement by a competitor.

It must be stressed that change will not manage itself smoothly and the natural tendency of managers to prefer a quiet life is counter productive.

Product generations

It is difficult enough to embrace change on a smooth management pathway. However, for many companies the situation seems even more difficult because the rate of change in their industry is accelerating.

A 'generation gap' trips up many companies in the faster-moving technological areas. There are two concurrent trends: first, product lives shorten. Thirty years ago, in many industries a newly launched product could expect a reasonably profitable life expectancy of twenty years. In the same industry today, a new product launch might have a life expectancy of only two years before it is obsolete. The second trend is that product development times are lengthening. Thirty years ago a two-year development period from idea to launch was

9

the norm in many industries; today the development time in many high technology industries is between ten and twenty years.

Together these two trends constitute a serious danger. It is all too easy to fall a whole product generation behind; and case studies indicate that companies which suffer this fate rarely survive. What we are saying therefore is that companies not only need to change and manage that process of change as smoothly as possible, but must manage the timing of individual business changes, that is, the development and launch of new business units, accurately if 'generation gaps' are to be avoided.

Profit shortfall

If a company wrongly times the introduction of its next generation of business, whether new products or new services, a shortfall against expected profits may occur. A 'profit shortfall' may seem no more significant than a temporary dip in the company's profitability. However, such a dip can have more profound consequences than first envisaged.

No company can be viewed in isolation from its competitive arena. A profit shortfall is a superb opportunity for your competitors to steal your business. If one of your competitors launches the next generation ahead of you, not only will you lose profitability in the short term but you are vulnerable to the loss of your client base in the medium to long term.

A generation gap is also a superb opportunity for a predator to acquire your company. That dip in the profits curve transiently devalues your assets and lays you open to acquisition as a cheap and potentially rewarding snippet.

On the other hand, it must be accepted that investment in innovation can produce a transient profit dip as the innovation absorbs current profits. Obviously, the trick is to find a cost-efficient approach to innovation.

To avoid these calamities companies must manage their innovation programmes consistently, continuously and in a smoothly accelerating curve. Anything less and there will be no status quo to maintain.

Vital need

We believe therefore that the ability to manage a company's innovation programme is vital to its survival. Yet we know, from our contacts with industry and commerce, that an average of less than one in a hundred companies has developed a cogent management system for innovation and many simply do not have a plan for innovation.

Moreover, we know that within virtually all companies there are manifestly powerful management groupings overtly dedicated to maintaining the status quo. When we ask whether boards of directors have evolved an approach to innovation they frequently misunderstand the question. Typically the answer is 'Yes, we have a strategic plan.' As we will see later, possessing strategy documents is insufficient to guarantee the company's survival. Innovation must go far beyond a paper plan – there are battles to be fought!

There is also a tendency at board level to assume that because a company has a research and development department the future is secure. This is also a fallacy. To believe that any single department can carry the responsibility for the company's future is naive. Rarely, however, do we find a company which recognizes that every department has a responsibility for the management of change.

The 1990s

In the fast-changing 1990s the target will be excellency in innovation. Whereas the 1970s and 1980s gave rise to many successful

company turn-rounds based on improvements in managing the status quo, for example by introducing sharper financial control and cutting out supernumeraries, we believe that 'company doctoring' in the 1990s will have to involve corporate psychology as well as corporate surgery. This is to suggest that rationalization will no longer be sufficient and a new range of management weapons will be needed to tackle the difficulties of innovation.

Concluding comments

What proportion of your departmental time resource do you spend fussing over transient trivia, and what proportion do you spend in detailed actions developing the next generation of profits?

11

3

Problems in business and industrial innovation

So if we must innovate, what problems can we expect? Many different pitfalls attend attempts within companies to innovate. In this chapter these are examined on a six-dimensional basis.

The question we are asking the reader is: 'Does your approach to innovation fall foul of any of these difficulties?'

Dimensional problem definition

Many companies know that something is wrong with their attempts to change and innovate, but fail to identify the particular cause. By defining business in terms of six separate but interacting dimensions we have developed an approach to pinpointing the causes of failed business and industrial innovation. The dimensions we use in our approaches to problem definition are:

1 *Sourcing* of ideas;
2 *Processing* and developing ideas;
3 *Marketing* and selling the new business concept;
4 *Timing* the idea generation and development processes;
5 *Financing* the research, development and launch processes;
6 *People* – psychology problems in innovation.

It will be seen that these are derived from the six primary dimensions outlined in Chapter 1.

By examining what has gone wrong in terms of these specific dimensions it is usually possible to pinpoint weaknesses in the current management of a company's innovation programme.

Sourcing difficulties

In any business or industrial sector there are companies which change only on the basis of 'me too' copying of their competitors' innovations. Clearly such companies have difficulties in finding, that is, sourcing, new business ideas internally and therefore have to copy externally generated innovations.

Examining such situations over the last decade has led us to believe that companies which have idea-sourcing difficulties are typified by the following characteristics.

1 *Authoritarianism* frequently stifles ideas. Because autocratic directors or managers over-exercise their authority no one else's ideas stand a chance. After a while everyone in the organization assumes all ideas for new business will come from the top. Unfortunately, autocratic people rarely seem to be 'ideas people' – they seem psychologically ill-equipped to be effective flexible idea generators.

2 *Departmental barriers* constrain ideas by denying otherwise available sources

within the company. A rigid departmental structure with little opportunity for multi-disciplinary contact creates walls through which ideas rarely penetrate. Cross-fertilization seems to be vital in idea sourcing.

3 *Rigid hierarchical structures* have a similar effect. Ideas become the preserve of the company's higher echelons and people lower down are positively discouraged from generating ideas. Yet there is little evidence to suggest that either seniority or experience guarantees ideas. On the contrary we will argue that over-dominant, unchallengeable expertise gets in the way of creativity.

4 Where *long tenure in a specific job* is the company norm, again new ideas are difficult to come by. A conventional wisdom pervades the organization and things are done in the traditional way. We wonder if the creative input to a fixed job declines linearly or even exponentially with time!

5 *Over-secretive companies,* in which people deliberately erect barriers between themselves and the outside world also have idea-sourcing difficulties. Because they are frightened of giving information away they lose the ability to communicate and collect new concepts or needs.

All such company structures have difficulties in sourcing new ideas for any form of change, particularly innovation. Blinkered approaches and fixed ideas stifle what creative flair there may be within their organizations. Without creative flair inside the company, their only choice is to copy competitors' innovations.

In our experience the archetypal example of companies with idea-sourcing problems are third generation family businesses. 'Clogs to clogs in three generations', we believe, has its root cause in the rigid hierarchical structure that exists in such companies. Although the founder of the company will usually have been very innovative, the second and particularly the third generation of the family tend to be backward looking, copy that which has gone before and aim to maintain the status quo.

A list of comments made within companies

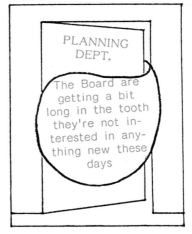

Figure 3.1 Idea sourcing difficulties

which we feel indicate innovation-sourcing problems is presented in Figure 3.1.

If readers have heard these or similar comments within their organizations, perhaps they are symptoms of internal innovation-sourcing difficulties. If you cannot come up with ideas in the first place, your innovation options are severely restricted.

Processing difficulties

Analysis of the causes of failed innovations within companies frequently identifies situations where a company has some very good ideas but finds it difficult to process them through to successful launch. As we will see later, this process is difficult and complex. To say that coming up with the idea is only the first part of the process is to state the obvious, yet many companies do seem to believe that ideas will launch themselves. Invention, the process of creating ideas, can only be the first and is often the easiest step in the innovation process. To be cost efficient in taking an idea through to launch a company must have considerable expertise in all the disciplines that will be involved.

Case studies in which venture development, that is, the processing of ideas, went wrong tend to indicate either a lack of expertise or an inability to appreciate what expertise is required. This situation has been exacerbated by the reduction in staffing levels which has taken place in most companies through the 1970s and 1980s. In management terms this has led to a situation in which there is no slack. Since for the most part the system is geared to managing the status quo, there is precious little available expertise to manage innovation. We believe it is a mistake to trim the management ranks to a point where the company cannot develop ventures for its own future.

We would also argue that the expertise required by a management team to manage the status quo is not the same as that required for the development of new business activities.

Analysis of failed innovation frequently exposes a company situation where the available expertise was imbalanced. If a company is lacking in one discipline involved in innovation then the process will fail. As we will see, venture development must be a multi-disciplinary approach involving 'marketeers' and 'profiteers' as well as 'technocrats', and it is a mistake to assume that one department can carry out all the functions involved in venture development.

Finally, we have many case studies which indicate that managements were naive in their initial estimation of the complexities involved in venture development, and this led them greatly to underestimate the required expertise. This gives rise to high hopes on day one, crisis management of concepts thereafter and, ultimately, wrongly blaming the manager of the function in which the venture failed.

Does your company have difficulties in processing ideas through to launch? The recorded comments shown in Figure 3.2 may be symptoms of such difficulties.

New business concepts do not launch themselves: they must be adequately resourced during their development, and appropriate expertise must be provided in all the disciplines of venture development.

Marketing difficulties

Our case studies strongly suggest, on a numerical basis at least, that misjudging the market is one of the commonest reasons for failed innovation.

In the 1970s and 1980s we collected many

Figure 3.2   Concept processing difficulties

case studies of failed industrial innovation which could be laid at the door of 'discovery push'. There were many spectacular failures in which companies discovered a new piece of technology which gave rise to a new product and then tried to force-fit the new product into an unwilling market. There are of course exceptions to every rule, and we have noted recently that several authors have again expounded the virtues of the discovery-push approach. We can say only that we do not believe the rare exceptions justify the risks involved.

Another way in which companies misjudge the market stems from their clients' conservatism. Few clients in business and industry will be interested in a new product or service simply because it is new. Indeed we know of many, particularly the older, industries in which new products have a tremendous barrier to climb before they will be accepted. The 'the devil you know...' attitude dictates that unless the new product or service has a specification far superior to the current norm it will not penetrate the market.

Another common reason for failure on the marketing dimension is not differentiating between influencers, buyers and beneficiaries. We believe this to be a targeting error which leads to much marketing effort being squandered. A classic example is the error of sending new product samples to the client's research and development department instead of the production department that indicated the need. Also we have numerous examples where companies tried to sell products which benefited an end user who did not pay for the product.

The recorded comments in Figure 3.3 should sound warnings in situations where marketing is failing.

As we will see later, there is a need for some very specific and highly targeted marketing techniques when dealing with innovations, and it may be that conventional market research is found lacking.

15

Figure 3.3   Concept marketing difficulties

Timing difficulties

Our case studies suggest that there are at least three ways of mistiming innovation.

The most obvious disaster is arriving at an already saturated market having taken too long to develop the product and thereby allowing time and space for competitors to get ahead.

However, it can be argued that it is equally damaging to arrive too early at the market place. Examples of products and services launched ahead of their time are legion. And the most galling situation of all arguably stems from an initially premature launch followed, after a period of withdrawal, by a relaunch which comes too late!

All these mistimings can be avoided, but only if you can accurately time the gestation period prior to launch and use this to work back to a timed estimation for a starting point for the innovation.

We have found that very few companies take product amortization seriously. They continue to push the same old product long after it has dipped below its profit plateau. Presumably they hope to milk the very last drop from their cash cows! However, this approach carries a danger: while maintaining this ageing product, competitors may well be giving birth to the next generation. And, given highly restricted company resources, you may find that maintaining the status quo on an ailing product devours all your resources leaving none for the newly born.

Unless a structured approach to idea generation is used by a company, ideas tend to occur haphazardly. This gives rise to a situation in which managements cannot predict the next generation of business.

Finally, we have a group of case studies in which the estimated gestation period proved to be a gross underestimate of the actual time required to take the concept through to launch. It is almost a joke in research and development departments in some companies that 'it will take three times as

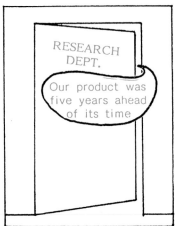

Figure 3.4   Timing difficulties

long as you think'. By this they mean that the uncertainties involved in venture development are bound to give rise to delays. Yet, as we shall see, techniques are available to help manage the various stages in concept and venture development. Using these techniques enables a far more realistic timing estimation to be made.

Does your company get its timing wrong? The recorded comments in Figure 3.4 are indicative of this difficulty.

With hindsight most managements can see the difficulty; the trick is to use techniques for checking timing as the venture develops.

Financial difficulties

Financing business innovation can be highly problematic especially where budgets are constrained.

The classic situations of innovation failing due to insufficient financial resourcing stem from the private entrepreneurial sector. Our involvement in the Science Park movement has given ample opportunity to witness these disasters. Private entrepreneurs and small firms and, on occasion, medium-sized companies can run out of money before the innovation reaches launch.

In many of these instances the initial estimate of the costs involved proved to be wildly inaccurate. One might coin the phrase 'it'll cost you five times as much as you think' to underline the order of inaccuracies found. In companies where the accounts department is paramount or where the finance director rules the roost we have found many exciting innovations starved to death. Indeed, we know of development departments in large national and even multinational companies where managers must juggle with financial allocations to obtain funds. A typical example is buying developmental materials and equipment disguised as spares for the mainline plant. Short-term, over-rigorous accountancy blights many a company's future.

As we will see later, finance is one of those

17

dimensions which can react perversely with other dimensions, for example time. In innovation there is a developmental progression which requires sequential financing, if finance and progression become out of step, the latter will break down. The cry of 'too little and too late' reverberates around development departments.

Another common pitfall is to fail to differentiate between a risky and a costly venture, to find oneself on a risky venture, then throw money at it only to find that the risk does not go away!

Still more tantalizing is the 'bicycle shed syndrome', in which a company will happily sanction a £2 million spend in the marketing department on an advertising campaign for a new product but will quibble over £1000 worth of uprated tooling for the same product.

We make no apologies for criticizing company accounts and finance departments which give rise to the type of recorded comments shown in Figure 3.5.

Over the last ten years the costs involved in business innovation have risen between 200

and 500 per cent depending on industry, yet we have many case studies from client companies where the spend on their future has actually reduced. We know of many industries which spend less than 1 per cent of today's turnover on their business for tomorrow.

At various points in this book we will argue that in the management of innovation a company needs a different psychology from the standard psychology of company management. We believe that many of the complex problems that defeat companies' attempts at innovation can be solved only by applying specific psychological techniques. Our multi-dimensional analytical approach is the first step in applying innovation psychology within a client company.

We believe that people who cannot or should not be involved in the innovation process display some characteristic traits.

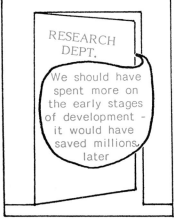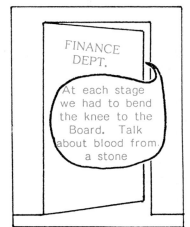

Figure 3.5   Financing difficulties

Figure 3.6   People difficulties

They have generally negative attitudes and whenever confronted by the need or opportunity for change their answer is 'No!' We christen them 'abominable no-men'.

No-men exist in all businesses and industrial companies and they sometimes gravitate to situations in which they can interfere with innovation. Most become troubled when drawn into this role whereas they are happy and efficient when employed to maintain the status quo. Ironically, when delving into the background of 'abominable no-men' we regularly find that they were once creative individuals, who have become disillusioned over the years.

To savour the people difficulties that are found in reality readers are invited to study the recorded comments in Figure 3.6. If you can spot the problematical dimensions you have a starting point for the application of some industrial psychology.

It is of course possible for several dimen-sions to be wrong at the same time! This gives rise to complex problems, many of which challenge our ability to understand what is going wrong. However, defining business in terms of six separate but interacting dimensions can be helpful in pinpointing the causes for failed business and industrial innovation even when failures are brought about by such complex interactions as:

- If both the timing and the sourcing are wrong you will become out of step with your industry. This amounts to an awareness problem.
- If you process the wrong resource in an innovation context, you will develop the wrong technology. This gives rise to inappropriate developments.
- If both the resourcing and the market are wrong you will have a focus problem. The focus of a business innovation programme should be in tune with the business mission of the company.
- If the market and financial parameters are wrong you will fail due to problems of cost effectiveness. Few markets are interested in innovation for its own sake

and new products must be perceived to be cost efficient by their intended market.

- If market and timing dimensions are wrong you will fail because the market will either be unready for your product or will regard it as obsolete.
- Problems with resource, process or marketing dimensions coupled with people problems will lead to failure through interfacial difficulties. The 'NIH (not invented here) syndrome' defeats many attempts at innovation.
- If people problems interface with the time dimension, delaying tactics will cause you to fail. New business development always causes some disruption, which must be managed and minimized if the innovation is to proceed to launch on time.
- With an adverse interaction between the finance dimension and time, your innovation will fail through cash flow problems.
- If there is a mismatch between specific dimensions, such as processing interacting with people, your innovation attempts will be bogged down by departmentalization problems.

- If people, time and finance dimensions interact adversely your innovation will fail due to commitment problems. A typical example of this is 'analysis paralysis'.

We would have to admit that multi-dimensional interactions challenge our modelling approaches to business, but in detailed studies of failed innovation attempts for client companies we have found that it is usually possible to shed some light on these complex problems by using our multi-dimensional approach. However, our studies of these complex interactions take us into the field of psychology, which in an innovation context is still very much an inexact science.

Try to work out which are the erring dimensions in the examples in Figure 3.7.

Summary

It is widely recognized that there are many different pitfalls along the innovation

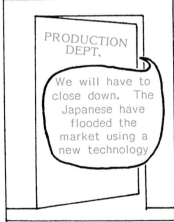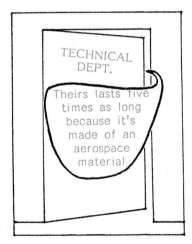

Figure 3.7   Complex difficulties

pathway. We believe that by using a multi-dimensional analytical approach it is possible to spot which dimensions are causing the difficulties in specific innovation attempts.

In the SWORD multi-dimensional analytical approach to innovation we believe there are six interacting dimensions:

- source
- process
- market
- time
- finance
- people.

In this context we do believe that people are a separate dimension; their input and interaction can be so problematical that they demand to be treated as a special dimension.

Using the SWORD analysis technique most companies' innovation problems can be appraised in terms of twenty-one permutations and combinations of the six dimensions. However, heaven help the company with all six dimensions wrong at the same time!

To check the validity of the SWORD multi-dimensional analytical approach, collect and appraise some apposite comments you hear about your company's innovation difficulties and analyse them on a six-dimensional basis, looking for both simple problems and complex interactions between the dimensions.

4

Approaches to successful innovation

In Part 1 we have so far defined innovation as the management of change in business and industry, stressed that innovation is vital in the rapidly changing world of technology and commerce in the 1990s, and pointed out that innovation is a difficult process with many pitfalls. In the last chapter we set out a multi-dimensional analytical approach for the study of innovation. In this chapter we take the recognition of these difficulties as our starting point, and go on to suggest that they are manageable but that they require specialist management and psychological systems.

The most usual problems

Having analysed the case histories collected during our consultancy activities it is possible to compile a list of the most usual problems which bedevil business and industrial innovation. Our database for this exercise extends to over 6000 failed ventures drawn from our conversations with over 1000 companies. (It is perhaps surprising that, on average, companies can cite six failed attempts at innovation – many cannot quote one success.)

Our analysis suggests that the three most common reasons for failure in innovation are:

1 *Marketing problems*
Typically companies say they had the market match wrong, that is, their new product or service did not fill the market niche planned.

2 *Interfacial problems (the NIH syndrome)*
Companies frequently complain that bad communications bedevilled ventures and ultimately led to their demise.

3 *Commitment problems*
These are exemplified by 'analysis paralysis' in the board – companies complain that the board's inability to take a decision involving risks led to the eclipse of the concern by a more dynamic competitor.

We believe that these three problems constitute the principal stumbling blocks to innovation in a majority of medium to large companies.

We should add, however, that small companies often suffer from more fundamental problems, most frequently financial, causing them to run out of cash before the new product or service reaches launch. However, one might argue that lack of focus or lack of appreciation for other component parts of the innovation process caused them to underestimate the necessary budget in the first place.

Problems and solutions

There are indeed many pitfalls along the pathway to business and industrial innovation, and they regularly defeat superb minds in highly successful companies. Hard-working, conscientious and efficient business managers frequently admit that they cannot guarantee the success of a new business activity. Our case studies show that they are powerless to prevent the crash of an erring innovative venture.

How is it that otherwise successful managers are unable to achieve the same consistent level of success in the management of change as they do in managing the status quo? Our answer is that the cost-efficient management of innovation is a completely different endeavour from the cost-efficient management of established business. We believe that the former requires an entirely different armoury of professional skills from those taught to practising managers for the day-to-day management of their companies. The day-to-day endeavour amounts to fire fighting to keep the established business on the road and, as such, the skills learnt for the management of ongoing business will tend to relate to the maintenance of the status quo, damage control, minimizing disruption and maximizing through-put of least diversity.

We believe that managers are not currently trained in the right management systems for the management of change. Moreover, we argue that innovation requires a different corporate psychology from the management of established business.

Innovative approaches

Over the past twenty-five years the author has been setting down, testing, perfecting and launching a range of special management techniques aimed specifically at handling business and industrial innovation. Out of this work has come the realization that a different psychological approach is required. Fortunately, in the final quarter of the twentieth century neurologists, psychologists and psychiatrists now understand the workings of the human brain sufficiently to provide a framework for its elemental understanding. This has been used to define a different psychological approach for use in innovation systems.

Neither these speciality management techniques nor the modified industrial psychology can be fully proven. Nevertheless, they do seem to have stood the test of time and they are offered on that basis. They may turn out to be very simplistic and clumsy approaches to highly complex human endeavours and no doubt they will be swept aside by more subtle systems in years to come. However, on the basis of clients already satisfied, new business already launched and improvements in corporate creativity already apparent, they are offered to the reader for trial.

Basic tenets

Eight basic tenets underpin all the systems presented in this book; they are shown in Figure 4.1 – can your company live with them?

These eight tenets constitute the battle plan we are presenting for innovation in the 1990s. If they seem reasonable ... read on!

Structure of Parts 2 to 7

Parts 2 to 7 of this book present programmes of specialist management systems for use at specific stages of the innovation process. Each programme is presented in four chapters, which follow the same sequence in each part:

1. **SYSTEMATIC APPROACH** - the innovative management systems presented are meant to provide a systematic approach to the various stages of the innovation process. As such, they will provide a systematic framework for the endeavour.

2. **MULTI-DISCIPLINARY APPROACH** - all of the systems presented are intended to be used in a multi-disciplinary situation. That is to suggest that all the primary disciplines within a company structure are to be involved in the innovation processes.

3. **NEEDS PULL** - the sequence set out in the systems concentrates firstly on recognising the problem or need in the situation, appraising this in the context of the means available to the company and then developing and launching a relevant opportunity. As such, the approach is sequential; "needs pull" first and then "discovery push".

4. **INTEGRATIVE APPROACH** - the systems are meant for the evolutionary management of change rather than coping with a revolution. They therefore concentrate on changes which can be integrated into the corporate structure and on developments which are incremental rather than disjointed.

5. **FOCUSED APPROACHES** - the techniques are all focused on specific business and industrial options. As such, they are targeted procedures rather than broad based theoretical approaches. This focused approach leads to specific systems.

6. **CREATIVE APPROACH** - underpinning each of the management systems presented here is a creative psychology. Throughout the aim is to build in a more open minded approach to management which in turn yields specialities rather than commodities because it focuses on novelties rather than conventionality.

7. **RAPID IMPACT** - the systems presented are intended to create a rapid impact on the innovation performance of a company. As such, they are not long haul management endeavours, their results will be apparent within months rather than years. In this way, they can be rapidly appraised and, if necessary, modified.

8. **ATTITUDINAL CHANGE** - both the systems and the underpinning psychology are intended to bring about attitudinal changes within corporate environments. Unless attitudes towards change can be changed, then no innovation systems will work.

Figure 4.1   Basic tenets of the SWORD approach to innovation

PART		
2	INNOVATIVE STRATEGIC PLANNING .. .."SPEAR"	
3	CREATIVE BUSINESS RE-ALIGNMENT .. ·. .. .."SABRE"	
4	NEW BUSINESS IDEA GENERATION .. .. .. "SCIMITAR"	
5	EXTERNAL SOURCING OF NEW BUSINESS .. .. .."BLADE"	
6	VENTURE DEVELOPMENT .. .. "RAPIER"	
7	NEW BUSINESS LAUNCH & COMMERCIALISATION .. "CUTLASS"	

Figure 4.2   The programmes in Parts 2 to 7

1   Problems.
2   System.
3   Psychology.
4   Format.

Programmes

The final chapter in each part presents a programme format. These programmes have been proved in the field on a consultancy basis. The six principal programmes which span the innovation process are set out in Figure 4.2. In these final chapters of each part we provide statistics and case studies on usage of our programmes.

Why the weapon names? We believe you need a veritable armoury to overcome the barriers to innovation.

PART 2

INNOVATIVE STRATEGIC PLANNING

STRATEGIC

PRODUCT

EVALUATION

AND

RESEARCH

SPEAR

Where do you start on innovation?

In Part 2 of the manual we present systems, psychology and a framework for innovative strategic planning which we believe should be the first practical exercise in a systematic approach to business and industrial innovation.

5

Problems in strategic planning

The need for a strategic plan

Companies have long recognized the need for a strategic or growth and development plan which should set out agreed business goals. Traditionally this has taken the form of a financial plan. Most companies bequeath strategic planning to their financial departments and ask them simply to extend the annual budgeting procedure to produce a financially based strategic plan covering perhaps three or five years ahead. Such a plan satisfies the need to be able to plan against a medium-term horizon in financial terms. It provides a pathway which is followed whenever strategic decisions must be taken on a financial basis.

However, we would argue that strategic planning decisions must be taken on many bases, not only financial, and therefore the strategic plan should be multi-dimensional.

Where is the company going?

Having said that most companies produce a strategic plan we should add that far less companies actually use one! Dare we imply that most company strategic plans, once completed, are filed away and forgotten until the end of the subsequent year? It is certainly true that most line managers are unaware of the detail of the company strategic plan. What tends to happen is that decisions are taken in line management on a fire-fighting basis and only at the year-end are the results compared against a strategic plan. If by chance they fit, good! If they do not fit, corrective action tends to be instigated at board level and handed down to line management. In this way line managers gradually learn the details of the strategic plan through their mistakes. The situation is at its worst in research and development, where it is rare indeed to find a formalized strategic plan. Yet in many companies research and development initiates projects for the company's future – about which the strategic plan is supposed to be written.

Thus, in many cases the company's financial controllers may believe that it is following the strategic plan, whereas the prime movers at line management level, for example research and development, may be moving tangentially, on the basis of initiatives taken in spite of the plan.

The limitations of conventional strategic planning

Even in companies where good communications prevail and line management are fully aware of the corporate strategic plan, there are difficulties. These stem from the nature of the conventional strategic plan: because it is formulated on a financial basis it tends to be uni-disciplinary or, as we

1. Theoretical Exercise

2. Poor link between Planners and Doers

3. Annual Myopia

4. Simple Extrapolation of Available Data

5. Filed under "Forgotten"

6. Used as an Instrument of Torture

7. Pious Hopes

8. Financial – uni-dimensional

9. Immovable

10. Insensitive to major changes

Figure 5.1   Problems with conventional strategic planning

would say, uni-dimensional. By this we mean such plans tend to deal only in financial matters with scant regard to either marketing or technological concerns. As such they cannot reflect change in circumstances either in the market place or in the company's field of technology.

Another difficulty stems from the way the plans are formulated by the financial department. Conventional strategic plans tend to be formulated looking backwards: past and current trends are plotted and then extended forward to the future. This simple extrapolation technique is prone to huge error in rapidly changing situations. On a financial basis the business might have assumed a smooth projectable curve, but changes in either market or technology which impinge on the business can totally disrupt the curve.

If significant changes occur on either the marketing or technological dimensions, it becomes virtually impossible to revert to the pathway of the strategic plan. Indeed, it can be disastrous to force-fit the changed company situation into an immutable strategic plan. The problems of conventional planning are shown in Figure 5.1.

The need for an alternative approach to strategic planning

We believe that the conventional form of strategic planning used by the majority of companies has been found lacking. Over the last five years we have therefore developed, launched and proven an alternative format. Known as SPEAR, this format is a multi-

1. Current state evaluation

2. Techno-commercial forecasting of current business

3. Amortisation Planning

4. Growth gap identification

5. Identification of integrative options

6. Identification of change options

Figure 5.2 The SPEAR approach to strategic planning

disciplinary approach to strategic planning. We argue that only through a multi-dimensional approach can changes on any of the dimensions of business be taken into account within the plan.

We also believe that strategic plans ought to be innovative rather than mere projections of historical data. Only in this way can they go beyond logical projections of historical data and take into account those significant and rapid changes within business and industry which go beyond logic.

Objective

The objective of SPEAR is to produce an innovative strategic plan which is owned by all departments in a company and which takes into account changes on all dimensions of business. This can then give rise to a corporate mission statement, which is understood and agreed by all line management. Through their involvement the managers of the company then have direct access to the formulation of the plan,

its updating and any required backtracking to the plan.

A subsidiary objective is to remove much of the secrecy and mystique which currently surround corporate strategic plans – the in-house secrecy and mystique! Although confidentiality must be maintained little purpose is served by the shroud of mystery which surrounds many companies' strategic planning.

Approach

The SPEAR approach is based on a form of three-dimensional business modelling. This approach provides for clear definition of the aspects set out in Figure 5.2.

As well as applying this modelling approach, there is necessarily a change in company psychology; this involves a move by senior management to a more open-minded, discussive stance. The psychological underpinning for SPEAR is presented in Chapter 7. First, however, let us examine the SPEAR model.

31

6

Strategic planning modelling system

In this chapter we present a format for the specific three-dimensional business modelling technique that we use as the centrepiece of the SPEAR approach to innovative strategic planning.

The purpose of the SPEAR model

By building a model of the company we create a microcosm of reality. This miniaturized version then acts as a convenient focusing device within which to spot trends and so on. The three-dimensional model gives far greater scope than traditional approaches used in strategic planning, which are either two-dimensional graphical techniques or purely financial uni-dimensional documents.

The model also helps to bring together all the departments in a company since it is both multi-dimensional and multi-disciplinary. As we have said, we believe that the company's strategic plan should be contributed to and owned by all departments of the company. The modelling approach helps to 'de-departmentalize' both the plan and the company.

Dimensions of the model

For strategic planning, the first step in the innovation process, we use three dimensions drawn from the six discussed in Part 1. The three dimensions used in SPEAR modelling are:

- products
- markets
- time

What of finance? The financial dimension features as a weighting factor within the model – read on.

It will be noted that products did not feature in the initial list of six dimensions. In our terminology products are processed resources and as such the product dimension of the SPEAR model is an amalgam of two dimensions: resources and processes.

Data collection

Once the axes of the SPEAR model have been defined the next step is to collect data on products, markets and time to set along the axes. All that is needed are two lists drawn up to show products sold and markets serviced on a year-by-year basis, both currently and in the past. This data is then set out along the three dimensions as shown in Figure 6.1.

All the products sold are set along one of the base axes and all the markets serviced along the other. The vertical axis is then

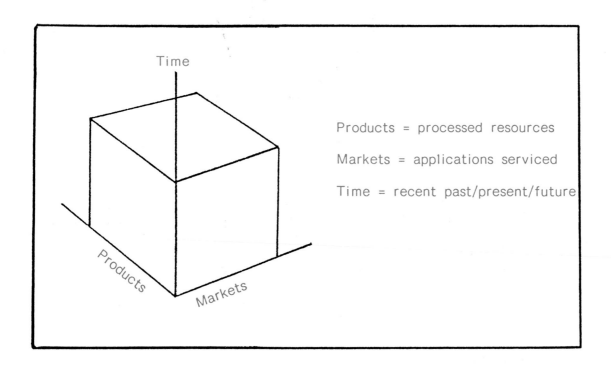

Products = processed resources

Markets = applications serviced

Time = recent past/present/future

Figure 6.1   The dimensions of a SPEAR model

made up of a sequence of time decks, typically ascribed to years. Business units are then defined as two-dimensional combinations on the base decks, that is, products sold into markets, and these are projected vertically where they lodge on individual time decks. The result is a sequence of columns of filled cells.

We normally build the physical model out of perspex sheets and small wooden cubes set out on a matrix scribed on to the sheets. In this form the wooden cubes represent business unit activity in a particular year. A column of cubes signifies an ongoing business activity in which the same product has been sold into the same market year on year.

It is convenient to ascribe other business data on the face of the cubes, for example the turnover, margin and profitability of each of the business unit activities. In this way one can see trends within the business

unit, defined on the faces of the cubes in the column.

Collation of data

To make the SPEAR model work, weighting data need to be collated on products and markets. These are the size indicators of business unit activity, for example turnover, margin and profitability. Product-weighting data is usually abundantly available in companies. Nevertheless, some skill is involved in defining individual products so that their model definitions are mutually exclusive. It is also necessary to differentiate between mark I, mark II mark III and so on of the same product.

Similarly, there is usually no difficulty in collecting information on markets. How-

33

ever, again some care needs to be taken on the definitions used to define specific market weightings so that they are differentiable within the model. Time will be required to separate specific information on specific products and on differentiating between the spread of markets served. The activity of separating out the profitability frequently exposes some highly unprofitable situations!

All these data must be collated on the basis of time, usually defined in terms of specific years. The company's financial year is more convenient than calendar years. Once the information is produced in this form it can be fed into the model, the individual business unit cubes carrying the data.

Checking the model

The model is checked by ensuring that the sum total of all business unit activities in a given year adds up to the turnover and profitability of the company in that year. A subsidiary check is made to see that no small and obscure products or markets have been missed.

At this stage the model will be a historical record of company performance leading through almost to the present day. (The current year may be virtually complete at the time of SPEAR modelling if the SPEAR exercise has been timed to coincide with the budgeting phase for the company. In such case the top deck of the model will at this stage be the current year.)

Extension

The next stage is to extend the model into the future. However, this is a great deal easier to say than to do!

As will be seen later, it is necessary to change the psychology of the situation to complete this step. Model extension into the future is carried out in two stages, the first by the company's executive team. They are asked to extrapolate current data and predict sales by product into markets in years to come, an activity similar to conventional strategic planning. It does have deficiencies and is therefore to be regarded only as the first stage in the extension process.

The second stage requires the input of an external agency. It also requires a far more open-minded attitude than most company boards exhibit, and we will be covering this ground in the next chapter. Suffice at this point to say that a second opinion on the future should be sought and then used to counterbalance in-house projections.

Opening the model search

Every cell in the model signifies a business unit in a specific time slot. Not all three-dimensional cells will be filled in any company's model because companies do not sell every product into every single market every year; there will therefore be voids. The filled cubes in the model represent the actual manufacture or procurement of a specific product and its sale into a specific market in the specific year. The faces of the cubes carry information on the size of the business unit activity, for example turnover, margin, market share and contribution. The filled cells are very informative, but what of the voids? It follows from the theory of the model that voids are inactive business units; in other words, they are products which were not sold into markets in that particular year. Not all voids in a company model will be meaningful because not every product can be sold into every market. Figure 6.2 shows a typical model.

The model is opened up for searching by focusing on 'the next year's deck', on which

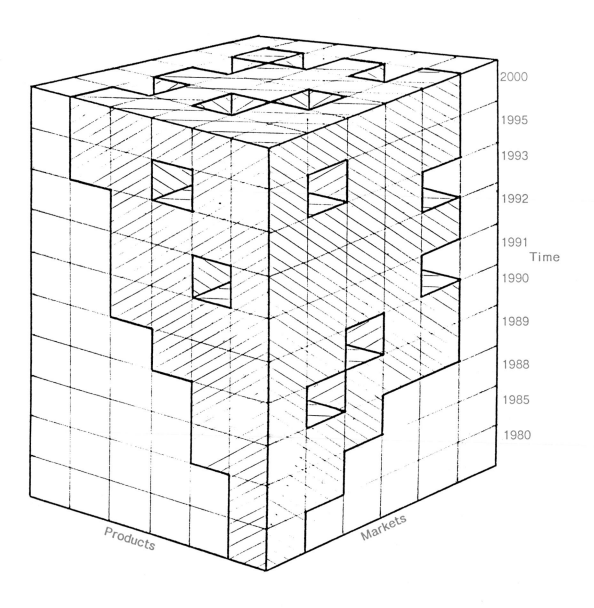

2000

1995

1993

1992

1991

Time

1990

1989

1988

1985

1980

Products

Markets

Figure 6.2   A typical SPEAR model

the array of filled cubes represents the company's projected business for next year. The size of the individual business units can be seen as weighting factors on the faces of the cubes. Conversely, all the voids on that deck represent combinations of business available to the company which are not perceived to be active next year. Not every void combination will be meaningful, as we have said, but they are all worth searching. By searching them relevant opportunities for new business will be found.

Available options

Three kinds of void will become apparent from searching the company's SPEAR model. These represent three available options.

The first option comprises those voids which represent business extensions. By studying the trends and the weighting factors carried on filled cells, that is, cubes in preceding decks of the model, it is possible to see changes in the company's performance *vis-à-vis* its existing products and markets. A study of these trends will yield opportunities for business extension. In particular it will focus on business units which are not doing very well or were abandoned and which could benefit from further management scrutiny and support. It is important at this stage to suspend judgement both as to why the trends are as depicted in the model and also as to whether the company can indeed sell more to reverse the trend. It is necessary only to write down the idea as a starting point at this stage. These business extension starting points are usually defined in the following form: 'There may be a possibility for selling (more of) product X into market Y in year Z.'

An opportunity of this type is often defined by a column of filled cells leading to a void on the current deck. Looking at the size of the business unit in preceding years it will be seen that this has dwindled to zero in the current year. This may well be a specific opportunity for business development – or it may amount to planned amortization.

The second form of business opportunity that arises from the early searches of the SPEAR model is 'gap opportunities'. These feature as a sequence of voids on decks leading up to and including the present. These can be defined as an opportunity but, again, you will need to suspend judgement. All that is necessary at this stage is to write down the opportunity: 'There may be an opportunity to sell product A into market B in year C – even though we have never sold it before.'

Perhaps the reader can see why it is necessary to change the psychology of the situation so as to trap all these embryonic opportunities. All too often the experts in a company will come up with ten good reasons why the opportunities are not real.

Development approaches

The third type of opportunity available in searching the SPEAR model is to look into those areas lying on the periphery of its current activities, which could be developed by the company.

Once the model is built in its initial form it is possible to see the company's current operational periphery. This is marked out by the furthest extensions along the product and marketing axes of the top deck in the model. Within this periphery all combinations are at least in theory available to a company; but what about the zones beyond? There will almost certainly be other products the company could offer and other markets the company could service. These possible new products and markets can be set out as further extensions to the model base deck.

When these are added to the model they generate new search areas, which should be explored to find new business opportunities of a developmental nature. Again it will be necessary to suspend judgement at this stage lest otherwise good ideas in a poorly defined form be discarded. The new search zones created by these extensions to the dimensions will give rise to new business ideas of the following type:

- 'Is this new product A capable of being sold into our existing market B in year C?'
- 'Is this new market X capable of being serviced by our existing product Y in year Z?'
- 'Is this combination of a new product R and new market S meaningful for year T?'

This last type of new business development is likely to be more risky than the other two because both dimensions are extensions rather than existing, but nevertheless it should be recorded.

Future projection

Having extended the base deck of the company to create search areas on the 'next year deck', we then use the SPEAR modelling exercise to project into the medium-term future.

This involves mounting new time decks on top of the 'next year deck'. Typically these decks represent two, three and five years hence (and on some occasions ten years). The additional decks are then to be populated with business projections for the future.

The easiest business projections are those which are continuous columns emanating from past decks. Voids are difficult to predict and peripheral voids even more so! Nevertheless, using the company's expertise and counterbalancing this with outside input it is possible to make some inspired guesses as to the business units available to the company in the future.

However, at this stage the SPEAR programme starts to move away from the systematic approach of the model into the psychology which we cover in the next chapter. Only by changing the psychology of the situation will the more nebulous concepts be discovered.

Concluding comments on SPEAR modelling

Three-dimensional modelling has been found to be a very useful technique in the context of strategic planning. The most useful three dimensions for strategic planning applications, drawn from the six dimensions of business, seem to be:

- products (processed resources)
- markets
- time

The financial dimension is implied as an underpinning of the product/market dimensions and is handled as a weighting factor on the filled cubes within the model. The financial implications of the strategic plan can be viewed alongside the product development and market development implications. In this way a multi-disciplinary approach can be brought to conventional strategic planning.

As we will see in the next chapter, it is also necessary to bring a new psychology to bear on conventional strategic planning as well as a multi-dimensional format, if realizable strategic plans are to be achieved.

7

Innovative psychology in strategic planning

In this chapter we present the modified psychological approaches which we believe should be built into a company's strategic planning exercise.

What has psychology to do with strategic planning?

As outlined in Chapter 5, strategic planning is a commonly abused management process in many businesses and industrial companies. It tends to be regarded by those involved as a chore and by those not involved as an abstract exercise. In Chapter 6 we argued that it should not be an abstract exercise and that all disciplines in the company should be involved in strategic planning.

In this chapter we will argue that strategic planning should not be a chore either. To be effective it should be a wholehearted management activity. And to be wholehearted it needs to be approached with a far more positive mental attitude than we have found in a majority of companies. In that such attitudes are psychological, so psychology has a part to play in strategic planning.

Current problems

Our quarrel with the psychology used in the preparation of a typical company strategic plan is that it settles for logic. This suggests that the plan of the company's future is based on logical extrapolations of the company's past and present situation.

In common with all managers in the more advanced West, strategic planners have been brought up on a daily diet of logic. From the age of five, education, training and management experience tends to reinforce the role of logic. Managers therefore try to see the logic in a situation and base all their conclusions on it.

But is the future of a company logical? And even if it is have we the capability of analysing the complexities of the situation to discover the logic? We have suggested that business is six-dimensional – we certainly believe it to be a very complex phenomenon. Within this multi-dimensional complexity things are constantly changing: not only individual dimensions, but businesses change through interactions between various dimensions. Can human logic unravel such a vast array of possibilities for the future?

We have seen many companies try to unravel these complexities to set down a cogent strategic plan. Few attempts in the 1960s, 1970s and 1980s have proved to be close to reality when the future arrived.

More recently we have seen attempts to use computer logic to define realistic strategic plans. As computer hardware and software advanced in the 1970s and early 1980s, many companies switched to a computer-based strategic plan. Yet again we have sincere doubts that the output approaches reality, and therefore question whether even the sophisticated and powerful logic of advanced computers is capable of 'seeing' into the future for a company.

Whether the logic of the strategic plan is based on the human brain or on advanced computers, the results seem to us to be weak extrapolations of known facts which fail to take into account rapid and perverse changes in the multi-dimensional complexities of business and industry.

extrapolation and yet something seemed to be wrong (if only that such a growth rate had never been matched by any industry since the early days of the Industrial Revolution). The history of Western steel industries since the early 1970s has indeed proven that there was something wrong!

In the 1990s the rate of business and industrial change will continue to accelerate, and a system must be found that goes beyond extrapolative logic. We will need lucky generals!

The need to go beyond logic

We therefore see a need to develop an input to strategic planning which goes beyond logic. Only in this way can a company take into account rapid and unexpected changes in its business; only in this way can a flexible, open-ended plan be formulated. One of the problems we see in logic-based strategic plans is that they tend to be 'yes/ no' affairs; they are very dogmatic, very fixed and much less open-ended. In the 1990s business will be changing so quickly that it will be vital to avoid dogmatism.

Many strategic planners and corporate executives feel that there must be something beyond the current logical approaches. One is reminded of Napoleon's famous comment when asked how he recruited his generals: his answer was that he recruited lucky generals! Add to that the quotation used by Richard Nixon, 'There are lies, damn lies and statistics!' On the one hand Napoleon seemed to be acknowledging that logic alone could not win battles whereas Nixon was underlining the view that, by their very nature, statistics can be woefully misleading. The author well remembers a heated debate in the early 1970s in a company servicing the steel industry, when a graph was presented suggesting a doubling in steel output per decade throughout the Western World. All the statistics presented supported this

Understanding the brain

To go beyond logic we turn to the fields of psychology, psychiatry and neurology. These sciences appear to be coming of age in the final quarter of the twentieth century and we believe they will be of paramount importance in the twenty-first century. Our understanding of the human brain is still far from complete, but sufficient knowledge is available to allow for some simple hypotheses to be checked in specific areas of endeavour. And there are simplifying hypotheses to take company boards beyond logic in formulating their strategic plan.

In the past five years we have used a sequence of very simple but workable hypotheses to develop a different psychology in the approach to strategic planning.

Most people in business and industry can appreciate that they have two ways of thinking about the future. For strategic planning purposes we use the hypothesis that people perceive things both 'at the front of the mind' and 'at the back of the mind'. We should make it clear immediately that we do not mean to imply neurological regions of the brain, rather that people feel that some of their perceptions are more frontal and dominant than others which seem to be subordinate and part forgotten.

We have chosen to use this hypothesis

because we have many times heard practising managers, involved in strategic planning, say things like: 'I've something at the back of my mind about this'. This 'something' at the back of the mind seems to us to be very important because it runs counter to the available statistics and logic of the situation. People feel intuitively that something is wrong about the frontal, dominant perception of what the future holds – a niggling doubt or fear that they have overlooked something which could fundamentally change the company's fortunes.

Time and again when we have managed to dig this fear out into the open, we have been able to counter a headlong pursuit of a goal which evaporated. In SPEAR we therefore use some psychological devices to create a mental environment in which these intuitive feelings are brought out and allowed to compete with logic.

From the back of the mind

Strategic planning is usually the preserve of senior executives in business and industry. Such people have a wealth of background experience and knowledge and should therefore be ideally placed to expound on the changes they have seen and both the successes and traumas that they have witnessed. This certainly gives them a sound basis for looking backwards at what has gone before. However, we feel they may need some help in looking into the future. It is all too easy for this wealth of information to be at the forefront of their minds as an all-pervading perception. Moreover, because companies tend to draw together a group of like-minded individuals with similar experiences, these perceptions will be self-reinforcing. They constitute what some psychologists call a 'mind set'.

We feel that such boards of directors need outside help in the form of an 'amiable amateur' to come in and help access the backs of their minds. From the back of the mind can come counter-currents, running against the mainstream of self-reinforcing logic. These vague notions challenge conventional thinking, that is, the company's mind set, and alert it to unexpected possibilities in the future.

All company boards are capable of avoiding mind sets and thus avoiding the business pitfalls of following preconceived notions to foregone conclusions. However, many boards require an external input to achieve this psychological breakthrough in strategic planning.

Framework

In SPEAR we provide a framework for both logical and intuitive strategic planning. This framework is the three-dimensional model outlined in Chapter 6.

The first stage of modelling is to model previous years leading to the current year. The second stage is to model future decks representing future years. The framework is ideally suited to logic-based extrapolations of available information and it is perfectly normal for boards to carry out this approach first.

Thereafter we insist that time is spent digging towards the back of the mind to find counter-predictions. And we insist that these are built into the strategic plan to make it more open-ended and less dogmatic. In this way we succeed in avoiding strategic planning becoming a chore which is done once a year and then filed under 'forgotten'.

System

Once the logic-based extrapolations of current business have been placed on future year decks the board is given a preamble about the workings of the human brain, focusing on 'front of the mind' and 'from

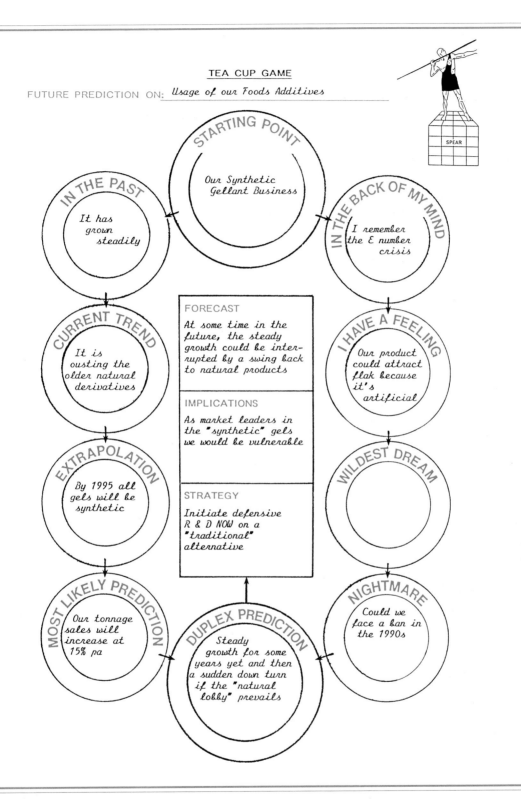

TEA CUP GAME

FUTURE PREDICTION ON: *Usage of our Foods Additives*

STARTING POINT — *Our Synthetic Gellant Business*

IN THE PAST — *It has grown steadily*

IN THE BACK OF MY MIND — *I remember the E number crisis*

CURRENT TREND — *It is ousting the older natural derivatives*

I HAVE A FEELING — *Our product could attract flak because it's artificial*

EXTRAPOLATION — *By 1995 all gels will be synthetic*

WILDEST DREAM

MOST LIKELY PREDICTION — *Our tonnage sales will increase at 15% pa*

DUPLEX PREDICTION — *Steady growth for some years yet and then a sudden down turn if the "natural lobby" prevails*

NIGHTMARE — *Could we face a ban in the 1990s*

FORECAST

At some time in the future, the steady growth could be interrupted by a swing back to natural products

IMPLICATIONS

As market leaders in the "synthetic" gels we would be vulnerable

STRATEGY

Initiate defensive R & D NOW on a "traditional" alternative

SPEAR

Figure 7.1   The tea cup game

the back of the mind'. Most board members acknowledge that although they cannot describe their feelings they do indeed recognize these undercurrents.

At this stage we then use a pro-forma sheet to dig out these contra-opinions and doubts. This sheet is deliberately couched in a light-hearted vein so as to reduce stress and anxiety. The aim is to get the board members to play the game as open mindedly as possible and to follow the sequence on the pro-forma sheet. As we will explain elsewhere in this book many of the psychological devices we use are drawn from children's games; this particular one harks back to 'consequences' and 'fortune telling', and is illustrated in Figure 7.1.

The aim of the 'Tea Cup Game' is to create time and space for these stray pieces of information and half-remembered perceptions to be brought to the fore to rival the dominant frontal logic of the situation.

Results

The result of the 'tea cup game' is a sequence of alternative perceptions of what the future might hold. It is most important that these results are faithfully recorded and presented alongside the logical extrapolations derived from the first stages of the SPEAR modelling procedure. There will be a natural tendency to believe the statistics and laugh at the alternative perceptions.

Once this second listing of strategic possibilities is logged, in the final debate at the end of the SPEAR programme both logic and counter-predictions can be

interwoven into an open-ended strategic plan.

Concluding comments

We believe that logic alone is insufficient to predict the future for businesses and industrial companies in the 1990s. Fortunately, from the fields of psychology, psychiatry and neurology, are coming workable hypotheses which help us to understand the human brain and to go beyond purely logical predictions.

We would not wish to claim that any of the psychological systems we use are new. Indeed their basis can be traced right back to Freud's work on the subconscious and to Jung on the power of imagination and intuition. All we believe we have done is to provide a non-threatening framework for company boards that stimulates this special kind of regression that we call creativity.

Within the accumulated wisdom, experience and knowledge of company boards there are, buried at the back of the mind, partly remembered perceptions and strange undercurrents. We recommend the use of simple psychological devices to dig out these counter-predictions so as to set them beside logical projections. In this way a far more flexible and open-ended strategic plan can result.

Many successful entrepreneurs appear to be very 'lucky' in identifying the need before anyone else. The conventional thinking and experiential learning of larger organizational structures takes the luck – and the success – out of many strategic plans! The non-threatening 'tea cup game' allows boards to spot the flaw in logical extrapolations.

8

A format for innovative strategic planning

In this chapter we present a format for both a systematic and creative approach to corporate strategic plans.

The SPEAR format

In Chapter 6 we presented a logic-based approach, that is, modelling, followed in Chapter 7 by a psychologically based useful 'game' to inject creativity into strategic planning. Each of these approaches can be used as a free-standing management system. In those two chapters we have tried to present our systems in simplistic forms so as not to constrain the way in which readers might apply them. You may try each, test it, modify it and use it on its own if you wish. However, we believe that in many company situations a more cost-effective approach is to apply these systems as part of an integrated format. Sword Innovation Services Limited has evolved such a format, which has been available as a consultancy programme during the last five years.

In this final chapter of Part 2 we set out the backbone of the SPEAR programme together with case studies and statistics that have arisen from its application in business and industry.

The programme

The format evolved for the SPEAR programme is a one-month exercise with the company board. A preliminary meeting of approximately half a day initiates the programme. This is followed two weeks later by the first full day of plenary meetings, and the programme is concluded two weeks thereafter with a final full-day meeting. The output from this final meeting is then presented back to the board in report form and constitutes an innovative strategic plan for the company.

Because board members are very busy, and indeed very expensive people, we have aimed for the minimum time commitment in the format which has evolved.

Model-based approaches

The aim of the preliminary meeting is to collect sufficient information from the board and from the company generally to be able to build the three-dimensional SPEAR model. Information is gathered from previous years' statistics on products and markets. Subsidiary information may also be necessary on technologies, resources and processes.

The three-dimensional model is constructed between the preliminary meeting and the first full day. This will have decks, each representing a previous year, and the top deck represents the current year. We have found it convenient to build the model both in the form of perspex decks with wooden cubes to represent the business activities, and using paper decks, which are easier to search.

Logic-based approaches

On the first full-day meeting logic-based approaches are used to project current trends into the future – the 'next year deck'. Techno-commercial forecasting techniques are used, and readers are referred to Chapter 5 *Industrial New Product Development* (see Bibliography).

These logical extrapolations are recorded for future comparison against contra-indications drawn from the creative approaches.

Creative approaches

During the afternoon the board is introduced to the 'tea cup game'. Between day 1 and day 2 the logical strand of tea cups is completed to represent the logic-based predictions of the future derived from techno-commercial forecasting techniques. The board is invited to consider possible contra-indications that the members will have at the 'back of their minds'.

Between day 1 and day 2 the external consultant produces a set of circular cards which can be shuffled and used down the right-hand side. These cards are meant to be mental triggers to awaken undercurrents at the back of the mind of board members on day 2.

On the morning of day 2 the tea cup game is played fully with the board. The aim is to fill in the garland of teacups on the right-hand side with provocative possibilities. These can then be cross-related to the logical extrapolations and this leads to a set of duplex predictions.

The duplex predictions are then discussed on the afternoon of the final day and become part of the innovative strategic plan. The 'amiable amateurs' need to

exercise considerable control on the afternoon of the second day, to prevent misinterpretation of the ideas. Only in this way can the company's mind sets be avoided and a creative strategic plan produced.

Statistics and case studies

We have used the SPEAR programme with a wide range of companies, in markets as different as computer peripherals and fast food, and have found it to be a powerful, thought provoking process which regularly breaks through mind sets within client companies.

Typically a SPEAR model for a company with an annual turnover of £25 million runs to forty product and thirty market sub-definitions, yielding a base deck of 1000 business activity combinations. We would then usually build ten decks in the vertical axis: five representing previous years, the current year's deck and four stretching into the future. Most companies prefer a deck for each of next year and two, three and five years ahead. Occasionally we have a deck representing a ten-year horizon. In this way the model generates 10000 cells. This is a far more detailed approach to the mechanics of strategic planning than most companies traditionally use. Typically it yields a hundred significant voids – listed in Figure 8.1.

Logic-based approaches are then used to search the model. While we have argued for processes that go beyond logic, we must emphasise that the logical approach is the first step. Many times we have discovered important strategic breakthroughs in logic-based approaches using techno-commercial forecasting. A typical example, drawn from the packaging industry, was the realization that simply by offering an own-branding service to all its clients as a matter of course, the company concerned could erect a significant barrier to entry against all its

SPEAR 100 CONCEPT PORTFOLIO FOR:

Presented below are the strategic planning options that we defined in the model building, systematic searching and creativity exercises during your SPEAR Programme

Current Business Options
1. 6.
2. 7.
3. 8.
4. 9.
5. 10.

Techno-commercial Forecast Options
1. 6.
2. 7.
3. 8.
4. 9.
5. 10.

Amortisation Plans
1. 6.
2. 7.
3. 8.
4. 9.
5. 10.

Growth Gaps Identified
1. 6.
2. 7.
3. 8.
4. 9.
5. 10.

Integrative Options
1. 6.
2. 7.
3. 8.
4. 9.
5. 10

New Product Options
1. 6.
2. 7.
3. 8.
4. 9.
5. 10.

New Market Options
1. 6.
2. 7.
3. 8.
4. 9.
5. 10.

New Presentation Options
1. 6.
2. 7.
3. 8.
4. 9.
5. 10.

New Forms of Business
1. 6.
2. 7.
3. 8.
4. 9.
5. 10.

Change Options
1. 6.
2. 7.
3. 8.
4. 9.
5. 10.

Figure 8.1   Typical output from SPEAR

> A low cost option that you could exploit immed-
> iately, is to offer, as standard and at no extra cost,
> to logo the inside surface of all the tubes you make
> for your clients, with your clients' house devices.
>
> This additional service would be valued by your
> clients who would not then go back to plain tubes
> bought from your competitors. Your competitors would
> not be prepared to speculate on logo dies for potential
> clients and could not therefore "poach" from you.

Figure 8.2 Systematic strategic option

competitors. This is nothing more than a logical approach and yet it was well received by the company's management. Figure 8.2 describes the option in detail.

By using the 'tea cup game' as a psychological device to open up boards of directors' thinking processes we have witnessed some remarkable contra-indications which have turned out to be very accurate. Typically a board will complete fifty tea cup games during the second day of SPEAR. A reasonable strike rate is five contra-indications which are so telling that they are built into the strategic plan. An example (slightly modified for security reasons) of the kind of output regularly achieved using the tea cup game is given in Figure 8.3.

A glimpse of the future?

An interesting spin-off from SPEAR is that companies frequently find themselves

> The prediction in this specific area of your busi-
> ness is that ecological pressure will continue to increase.
> This will threaten all of your existing ranges including
> your current "cleaned up" versions. (Ecologists will
> argue against aromatic products in principle). We
> therefore fear that your current research is pouring
> "good money after bad".
>
> The creative option that arises is the possibility
> of finding an aliphatic alternative with similar effective-
> ness - and far less ecological danger.

Figure 8.3 Creative input to strategic plan

Business computers will be a rarity by 1995.

The last oil shortage will be seen to have been 1974.

Vast desert areas will be recultivated by 2000, changing the World's geography.

Metallurgical production will be sea-borne by 1998.

Most commuters will wear masks to work by 1992.

Surgery will be a thing of the past by 2005.

International Travel will decline from 1995.

Gold will be as cheap and plentiful as copper by 2010.

Figure 8.4 SPEAR predictions

predicting the future for techno-commercial endeavours which are peripheral to themselves but which could have a significant impact, through technology transfer, to other industries. As such although our clients found them intriguing they could not be acted upon. We list in Figure 8.4 a few such predictions; they look odd, certainly they are not logical, but could they be real?

Output

The output from a SPEAR one-month programme is a six-part strategic plan, each part containing five investigational proposals, as outlined in Figure 8.5.

Concluding comments

The one-month SPEAR programme involves the whole company board in the preparation of an innovative strategic plan. Conventional thinking and simple projections of current data are challenged. Important strategic options are regularly discovered and can be actioned before competitors know what hit them!

Can you perceive the battles that will be fought over your customer base in the mid-1990s?

SPEAR FINAL PORTFOLIO for			
The following Strategic Options for Investigation emerged after assessment and selection of the initial portfolio.			
Approach	Sector	Option	
Current Business			
Techno-commercial trends			
Amortisation Plans			
Growth Gaps Identified			
Integration Options			
Creative Change Options			

Figure 8.5   SPEAR strategic plans

PART 3

CREATIVE REALIGNMENT OF BUSINESS

SPATIAL
  ANALYSIS
    BUSINESS
      REALIGNMENT
        EXERCISE

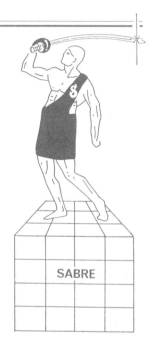

'So, we came up with all these creative strategies to investigate, what do we do then – we've got no one available to work on them?'

In Part 3 we cover the second step in our systematic approach to business or industrial innovation. Once an innovative strategic plan is in place the second stage can be addressed. We recognize that the second stage must be a creative realignment of existing business so as to free management resource for practical innovation.

9

The need to realign business

In this chapter we set out the basic concepts of our approach to business realignment, which we hope is more creative than the average 'night of the long knives'! And we do mean realignment, not rationalization.

Strategic needs

New ventures will not launch themselves: resources must be found to execute an innovative strategic plan.

Many companies have slimmed down to such a point that they have no spare capacity, that is, human resources, to carry forward any new ventures. It is normal to find at the end of the strategic planning exercise that companies lack sufficient resource to execute chosen strategies. Yet it is very difficult to argue for an increase in the headcount to develop profits in the future. Although companies tacitly accept that they need to speculate to accumulate, it is often very difficult to generate additional resources in a real situation.

It is possible to subcontract development activities and use external agencies to develop new products or services in line with the company's strategic plan. While there has been considerable growth in such activity in recent years, most companies prefer to develop their own ventures in-house if at all possible. The solution is therefore to find some resources in the slimmed-down company situation which can be switched to investigating strategic ventures.

The need for realignment

The alternative is to realign management resource and time in such a way as to free individuals for venture investigations. This is not to be confused with a rationalization process to reduce the headcount. What is to be achieved is to switch staff from ongoing duties, which can be shelved, to the development of future ventures, which should not be.

Most companies doubt that, after rationalization, there are any opportunities for this kind of switching; indeed, if a purely systematic search of the organizational structure is carried out very few opportunities will arise. Nevertheless we have found that by using a more creative approach it is possible to find manpower and management resources from within an existing structure, without any detriment to the ongoing business.

More creative approaches

The programme that follows SPEAR is SABRE. This is a more creative approach to the realignment of available manpower resources than is usual within companies. SABRE is a team exercise in which the members are drawn from all the principal disciplines and departments within a company. Its aim is to free resource and manpower for strategic developments. The SABRE approach is to model the company in its current state, compare and contrast resourcing and manning levels in various areas within the three-dimensional format and seek out under-utilized resources. Then, by the realignment of duties it is always possible to make time available for some key personnel to carry forward the strategic investigations identified in SPEAR.

The approach

The SABRE approach is another form of three-dimensional company modelling which is searched creatively.

Perhaps the simplest definition is to say that in SABRE we try to aggregate and equate workload across departments and individuals. In most companies the workload is significantly greater for some than for others. Conversely, if one looks closely enough there are some individuals who can be regarded as under-utilized resources.

The aim, of course, is to free key individuals. This must not, however, threaten current business.

Searching

Once built the SABRE model is searched in two stages, systematically and then creatively. The systematic search may be expected to generate a few fairly obvious but nevertheless worthwhile savings. The creative search may be expected to go beyond the logic of the situation and engender a new perspective in terms of human resource management. Out of this creative search will come some far-reaching realignments.

10

Realignment modelling

SABRE modelling is another format within which we use three of the six dimensions of business to produce a three-dimensional search area. This model is used to find savings, realignment and resources for venture development.

The aim of SABRE modelling

The aim of SABRE modelling is to free a multi-disciplinary team for innovation activities from within the company's existing organizational structure. In this way innovation activities can be made very cost efficient.

Why a multi-disciplinary team? We do not believe it is sufficient to have a research and development department and to charge it with the sole responsibility for innovation. Such departments have, in our experience, far too limited an experiential base to carry ventures through to launch in a cost-efficient way. We believe that a multi-disciplinary team approach improves communication within a company, assists collaboration between departments and spreads knowledge through the organizational structure. For these reasons we choose a multi-disciplinary approach to innovation, wherever possible.

To achieve this it is necessary to second marketeers, profiteers and technocrats into a SABRE team. It can be argued that companies which do not have a research

and development department find this step easier than those who do!

The axes

The axes we have chosen for the SABRE model are derived from the theoretical dimensions of business outlined in Part 1. The three chosen on this occasion derive from the three primary dimensions:

- source
- processes
- market

However, as is often the case in highly specific models, we have found it necessary to develop and use derivative dimensions rather than these three primaries. The three derivative dimensions normally used in SABRE models are:

- management task
- management option
- business unit application

These three axes are set out along the three dimensions of a business model of the client company, as in Figure 10.1.

Data collection

Data are then collected on the three derived dimensions and ascribed along the three

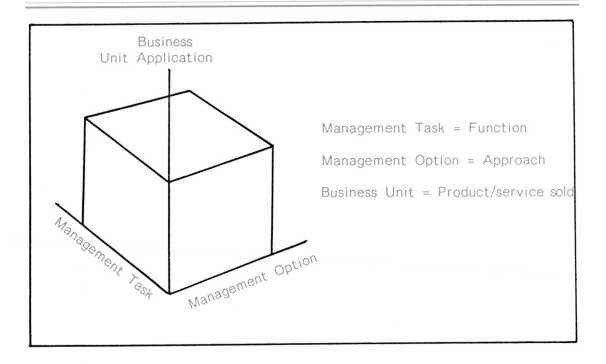

Figure 10.1   The dimensions of a SABRE model

available axes.

The first task is to redefine the roles of all the individuals and departments within the management of the company. The easiest way to do this is to subdivide all the roles into primary groups:

- management and control
- management of human resources
- developmental management
- production management
- marketing management
- financial management

Within each there will then be specific roles. A typical array for the management role axis of a SABRE model is shown in Figure 10.2.

The next dimension to tackle is management options. Again, the easy way is to subdivide this axis into groupings of options first, and then break open each group of options. Typically the grouped

options in a company's SABRE model are:

- rationalization
- automation
- agglomoration
- contracting out
- abandonment

These grouped options (also shown in Figure 10.2) constitute the ways in which management could perform the tasks set out along the role axis.

The third axis of the SABRE model is concerned with the business unit activities of the company. Again, it is usually easier to group these first and break down the groups later. Typical groupings along this axis will be the divisions of a company. Each deck in the model then represents a division.

At a later stage each division can be broken down into individual business unit activities. A business unit activity is the

MANAGEMENT TASK (FUNCTION)	MANAGEMENT OPTION	BUSINESS UNIT
1. General Management	1. Leave unchanged (As is)	1. Product X Component Manufacture
2. Planning	2. Minor trimming	2. Product X Sub-assembly
3. Information systems	3. Major surgery	3. Product X Assembly
4. Production scheduling	4. Discontinue function	4. Product X Installation
5. Purchasing	5. Automate	5. Product Y Circuitry Production
6. Goods inwards	6. Computerise	6. Product Y Boards
7. Raw material inspection	7. Contract out - part	7. Product Y Sub assembly
8. Stores	8. Contract out - entirely	8. Product Y Peripherals
9. Issuing	9. Spin off	9. Product Y Fitting Out
10. Component manufacture	10. Buy in	10. Product Z Old Format
11. Assembly	11. Amalgamate (2 units)	11. Product Z Type 2 Format
12. Testing & QC	12. Combine (more than 2 units)	12. Product Z Development Format
13. Process control	13. Centralise (all units)	
14. Packing	14. Develop	
15. Despatching	15. Expand	
16. Invoicing	16. Re-open	
17. Credit control	17. Relocate	
18. Stock control	18. Regionalise	
19. Auditing	19. Integrate	
20. Corrective action	20. Sell off	
21. Repair	21. Management buy out	
22. Improvements	22. Close down	
23. Maintenance		
24. Recruitment & training		
25. Health & Safety		
26. Industrial Relations		
27. Pay/salary		
28. Performance control		
29. Security		
30. Cleaning		
31. Technical Department		
32. Market Research		

Figure 10.2   Typical SABRE model

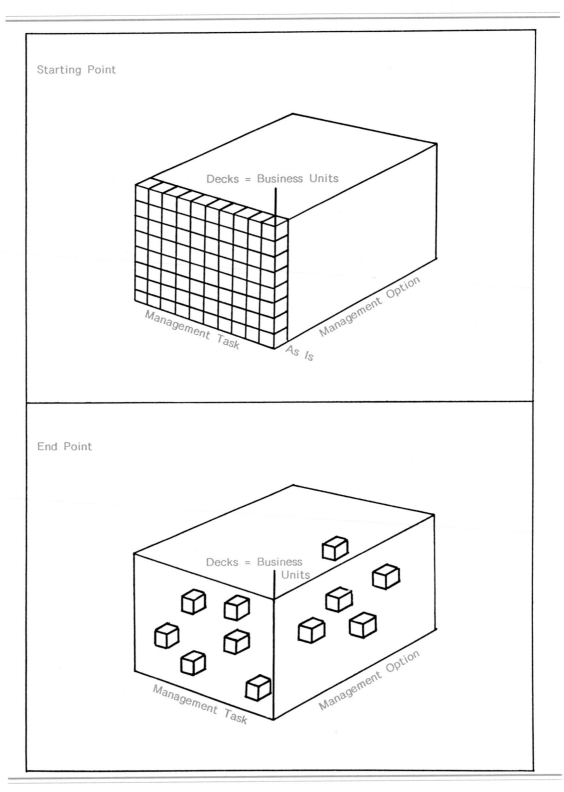

Figure 10.3   SABRE model

SABRE Model — Deck: (Product) Component Manufacture for P29 — X: TASK	Y: OPTION	1 Leave as is	2 Minor Trim	3 Signif. Cut	4 Major Surgery	5 Electric Autom	6 El'tronic Auto	7 Robotise	8 Flexible Mfr	9 Part Contr. In	10 Compl.Contr.In	11 Spin Off	12 Part Buy Out	13 Compl. Buy Out	14 Amalg. on Site	15 Reloc. on Site	16 Centralise on s.	17 Amalgam.Sites	18 Relocate Site	19 Centralise Site	20 Amal.Nation'ly	21 Reloc.Nation'ly	22 Cent'lise Nat'ly	23 Sell Off	24 Close down	25	26	27	28	29	30
Gen. Management	1	1																													
Planning	2	1																													
Info. System	3								3																						
Product'n Sched.	4						6																								
Purchasing	5					7																									
Raw Mat. Inspect.	6				2																										
Stores/Issuing	7					7																									
Component Mfr.	8								8																						
Assembly	9								4																						
Testing & QC	10	1																													
Process Control	11				5																										
Packaging	12		9																												
Despatch	13			10																											
Invoicing	14		11																												
Credit Control	15		11																												
Stock Control	16		12																												
Auditing	17		11																												
Cost Control	18		13																												
Maintenance	19			14																											
Resource Plan'g	20				15																										
Recruitment	21		16																												
Training	22		17																												
Health & Safety	23		18																												
Ind. Relations	24				19																										
Staff Develop.	25								20	21	22																				
Pay/Salary	26																														
Perform. Control	27																														
Security	28																														
Temp. Labour	29																														
Cleaning	30																														
Technical Dev.	31																														
Market Research	32																														
Budgeting	33																														
Research	34																														
Financial Anal.	35																														
	36																														
	37																														
	38																														
	39																														
	40																														

Figure 10.4   SABRE chequer sheet

manufacture of a specific product range for sale into a specific market sector. This range of business unit activities then constitutes the third and final dimension of the SABRE model. Again, these are shown in Figure 10.2.

Model building

Once all the necessary data have been collected from within the company the model can be built.

We usually use perspex sheets and wooden blocks. Traditionally the vertical axis in SABRE models is reserved for the individual business unit activities. In this way each business unit activity constitutes a deck in the model. Along the two axes of these decks are ascribed all the management roles and management options. At the beginning of the SABRE exercise all the first-ranking cells are filled, since at the beginning no realignment has taken place and therefore all the roles are being achieved 'as is'.

When completed in its initial form the SABRE model has a complete banked wall of filled cubes in the 'as is' column running all the way down the axis of management roles and all the way up through all the business unit activities, as in Figure 10.3.

Systematic searching

The model is now ready for searching. The first-stage search is purely systematic and is achieved using a chequer-sheet approach. Paper sheets representing individual decks in the model are drawn up as shown in Figure 10.4.

The first-ranking position – the 'as is' column – is currently filled. Chequer sheets provide an opportunity for systematically searching all the void combinations corresponding to other ways of achieving individual management roles. These range from minor reductions in resourcing to the abandonment of the function or role. Chequer sheets are fairly challenging in the SABRE exercise and management need to be encouraged to see the positive side: freeing manpower and resources to tackle the developmental future of the company.

At the end of the systematic search of all the available cells in a SABRE model (running into many thousands) several options will be found. It is necessary, however, to persuade managers searching the SABRE model to suspend judgement at this early stage as to whether the consequences of the proposed move are viable.

11

Creative change options

In this chapter we move from a systematic search of the SABRE model to a creative search. We outline the need for a different psychology in terms of personnel management and then present a specific creativity technique which we have found helpful.

Digging a little deeper

In most SABRE exercises we are looking to free a small percentage of the available manpower and resources of a company for venture activities. As we have said, SABRE is not intended as a pure rationalization exercise. Nevertheless, some companies have used it to generate a considerable pool of realignable manpower and resources. In either case it is necessary to 'dig a little deeper'.

We have tested many creativity techniques in SABRE and have settled on the following 'Playing Card Game' approach. Before introducing this, readers may find it useful to understand the rationale of the game.

Staring into the abyss

When looking for further realignment opportunities within a SABRE model we are really asking company management to stare into the abyss. We are discussing their people in their teams and their departments. Faced head on this can be a highly stressful experience. The human brain dislikes staring into an abyss but can often cope with it if it is disguised. The 'playing card game' provides this disguise.

The playing card game

The playing card game involves labelling standard packs of playing cards to create three sets. The first set constitutes disguised scenarios and the second set constitutes disguised options. The real options are on the third set.

The disguised scenarios are drawn from the world of nature plus human situations. The aim is to choose mildly amusing disguises for real-life painful situations. To give a flavour of the kind of disguises used successfully in the game, some examples are set out in Figure 11.1. The disguises are typed out on labels and stuck to the backs of the first pack of cards.

The face of the first pack of cards contains lists of, usually five, possible problems that could befall the scenario on the back of the cards. Again, these 'problems' are best explained by example – see Figure 11.2.

So, the back of this first pack of cards contains some analogies and the front some problems that could face those analogous

Instructions: Pick three cards that relate to your business from the first pack – set out face downwards

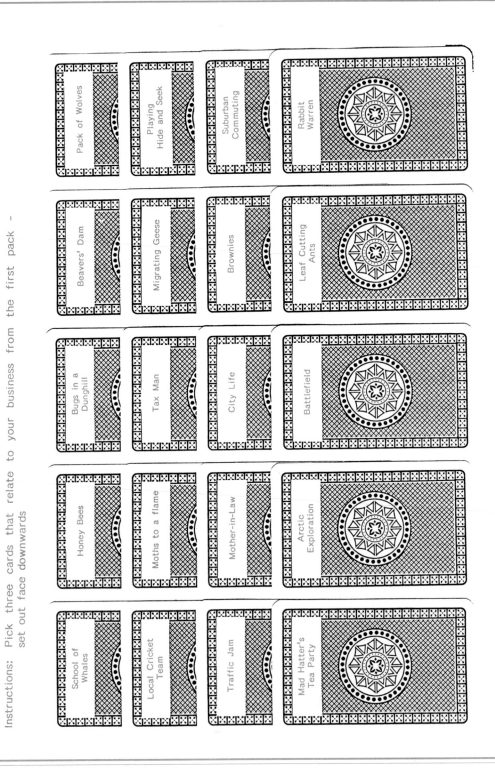

Figure 11.1   Playing card game – disguised scenarios

Instructions: Turn the three cards you have picked and read the analogous problems. Pick the most poignant from each card.

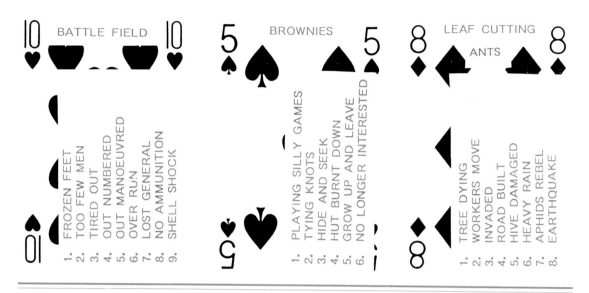

Figure 11.2   Playing card game – analogous problems

Instructions: Pick out the equivalent card (by suit and number) from the second pack. Look for a possible solution to your chosen analogous problem and note the code letters indicated.

Figure 11.3   Playing card game – disguised options

Instructions: From the third pack, pick out the words carrying the letters you chose, turn and read the real options.

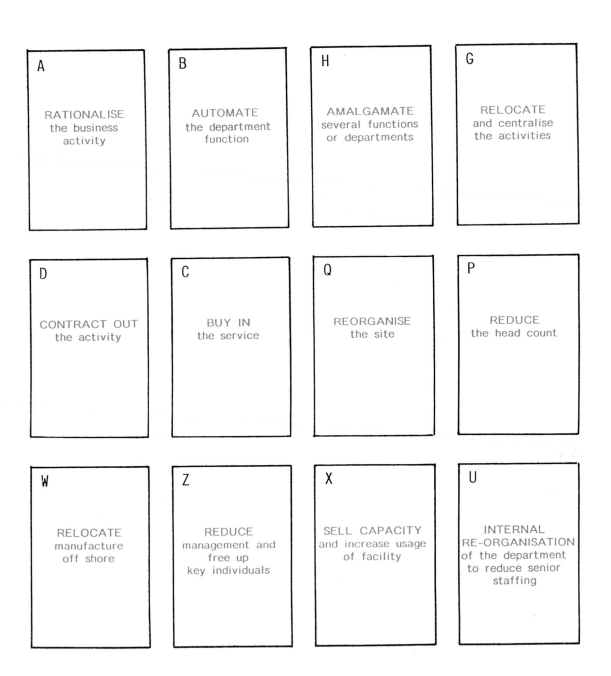

| A | B | H | G |
| RATIONALISE the business activity | AUTOMATE the department function | AMALGAMATE several functions or departments | RELOCATE and centralise the activities |

| D | C | Q | P |
| CONTRACT OUT the activity | BUY IN the service | REORGANISE the site | REDUCE the head count |

| W | Z | X | U |
| RELOCATE manufacture off shore | REDUCE management and free up key individuals | SELL CAPACITY and increase usage of facility | INTERNAL RE-ORGANISATION of the department to reduce senior staffing |

Figure 11.4   Playing card game – real options

CREATIVE IDEA GENERATION SEARCH

Site: S.E. _____   Production Line: GMs

SUSPEND JUDGEMENT

ANALOGY	PROBLEM	POSSIBLE SOLUTIONS TO ANALAGOUS PROBLEM	OPTION KEY	OPTION INDICATED	POSSIBLE INTERPRETATION (IDEA)
MIGRATING GEESE	THUNDER & LIGHTNING	GO TO GROUND	A	RATIONALISE	DEVELOP TYPE RANGE
			P	DECREASE	SELL CAPACITY
			X	SELL CAPACITY	& BRING ABOUT A TOTAL SELF-CONTAINED BUSINESS UNIT
BEAVERS DAM	POACHING	DIVE DEEP	A	RATIONALISE	BASED ON S.E. INFORMATION SERVICES
			J	CLOSE	FREE UP
			X	SELL CAPACITY	2 TECHNOCRATS
MOTHER IN LAW	TITTLE TATTLE	USE HER AS A SPY	E	BUY IN	1 MARKETEER
			I	DEVELOP	1 PROFITEER
			Y	TOTAL BUSI-NESS OPTIONS	

Figure 11.5   Playing card game progress sheets

situations, and these have a tentative link to the real problems facing the real company. (Skill is required in picking the analogies.)

The second pack of cards is the option pack. On the face of the option cards there are some possible solutions that could be applied in the analogous situation to solve the problems. Examples of these possible solutions are set out in Figure 11.3.

Finally, on the third pack of cards we set out the real options. Again, some examples are presented in Figure 11.4.

Playing the game

To play the game departmental managers follow options by key letters. These key letters take the player through a mental sequence. The aim of the exercise is to make that mental sequence less painful than would otherwise have been the case. In the playing card game we monitor managers' progress on progress sheets, as shown in Figure 11.5.

By playing the game in their spare time over a couple of weeks, managers within companies amass a second portfolio of realignment options. These are usually far more creative and interesting than those generated by the purely systematic search carried out on the chequer sheets.

12

A format for the creative realignment of business

In this chapter we present a format for in-company creative realignment. Both SABRE modelling and the card game can be used as free-standing management systems. However, we have found that companies benefit from a carefully formatted programmed approach, which is outlined in this chapter.

The SABRE format

Both the modelling approach outlined in Chapter 10 and the 'playing card' creativity system outlined in Chapter 11 can be modified to fit specific management situations, and we invite Readers to test these basic approaches in their organizations. Many companies have successfully applied SABRE modelling on their own so as to free people and time for innovation. Others have used SABRE modelling for straightforward business rationalization; however, this is not its intended application. Certainly, the psychological approaches used in the card game are more appropriate to the creative realignment of business in preparation for a concerted innovation programme. The SABRE concept may be used as a consultancy programme, the format of which is outlined below.

The programme

The format evolved for the SABRE programme is a three-month exercise involving departmental heads within the company. The programme commences with an introductory meeting attended by both board and departmental managers at which an outline of the SABRE programme is given and during which the real need – venture resource creation – is explained. This discussion leads to confirmation of the SABRE team of departmental managers. This is itself an important team-building exercise (readers are referred to Chapter 4 of *Industrial New Product Development* –see Bibliography).

About a fortnight later, day 1 proper of the SABRE exercise takes place. At this meeting the dimensions of the model are established and the current resource utilization (the 'as is' position) is defined as a number of weighting factors. The information presented is then assimilated and collated into model form. This process takes between two and three weeks, after which a second full-day meeting is convened.

On day 2 of the SABRE programme the model is built and the team is invited to study the current resource allocation within the company. Frequently this gives team members a new perspective on their company, its operations and the costs/benefits associated with individual business lines. When first built, all the filled cubes in the SABRE model are on the first rank, that is, 'as is'. The current resource allocations are carried as weightings on the cube faces. The team is then invited to contemplate moving the filled cubes along the realignment axis, that is, the management options dimension.

Between day 2 and day 3, approximately three weeks, the team works on a systematic approach to logical management options for the realignment of resources and business. On day 3 these systematic approaches are tabled by the team.

On the afternoon of day 3 the need to inject creativity in the exercise is stressed to the team and the 'card game' is introduced. As stated earlier the aim of the game is to make otherwise unpalatable options acceptable to the team.

Between day 3 and day 4 the team use the card game to generate a second and more creative portfolio of management options, which are presented on day 4. There is often a very dynamic discussion in which cross-fertilization of individual management options takes place. The end point of day 4 is the completion of the second portfolio of management options.

At this stage the weighting factors, that is, the current costs and benefits of the existing organizational structure, are fed back into the equation (this information was collected on day 1). The team then goes away and costs out the preferred management options generated during the SABRE programme in terms of cost savings, resource savings and impact on current business. (By withholding these weighting factors until all the ideas are safely recorded we avoid pre-judgement.)

On day 5, the final day of the SABRE programme, each team member presents a short-list of preferred options for business realignment. These options are fully costed in terms of the benefits and savings they will achieve.

This interlocking approach to the various stages of a business realignment exercise has been shown to be highly cost efficient in terms of the saving and freeing of resources in preparation for an innovation exercise.

66

SABRE models

The average size of a SABRE model is 25000 cells. However, it is not unusual to find that larger organizational structures generate models of between 250000 and one million cells. Within these enormous models there are always a multiplicity of business realignment options.

A typical systematic option portfolio generated from a SABRE model will approach 150 specific management actions. Equated to an average company turnover of £25 million a year these systematic options imply a saving in excess of £1 million a year in operational costs, or a 5 per cent freeing of resources for an innovation programme.

The systematic options drawn from a SABRE model are often fairly obvious once written down. However, it is in the nature of things that until they are written down they are never discovered and the effect of the SABRE model is to bring these fairly obvious options to the fore. Statistical outputs from SABRE are presented in Figure 12.1.

Creativity output

A typical output from the card game is a further 300 management options. These options frequently amount to a significant percentage of the total operational costs of a company – in extreme circumstances their impact can result in as much as a 25 per cent saving on establishment costs!

The types of ideas which come out of the creativity exercises within SPEAR are far more radical than those derived from the systematic approach. We believe that many of the options would never be voiced if it were not for the palliative of the card game. This allows managements to perceive and

Average output from SABRE Programme

Model   40 Management Activities)
        30 Management Options   ) 24,000 cells
        20 Business Units   )

Initial Portfolio

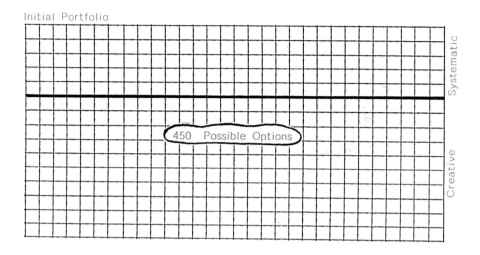

First Stage Appraisal

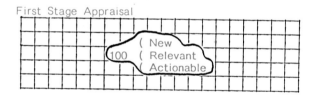

Second Stage Appraisal

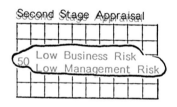

Third Stage Appraisal
    - 10 High Yield to Cost Ratio

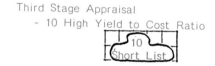

Short List of Viable Ventures for Business Realignment

Figure 12.1   Statistical output from SABRE

discuss potentially painful options in an open-minded and free-thinking way.

It is not unusual in the second-stage search of SABRE – the card game – to produce a single management option which is worth several million pounds in saved operating costs.

Case studies

Whilst it would obviously be wrong to quote detailed case studies of specific companies, the generalized example in Figure 12.2 should suffice.

A major electronics company used SABRE in 1987 and modelled all its UK operations. The model was searched by individual site directors and each produced a short-list of 12 possible options:

* 4 local options

* 4 cross-site options

* 4 radical options

The total implied saving of these options exceeded £20 million and freed up over 300 individuals, many of whom could be re-aligned to perform innovation and change functions within the organisation.

Figure 12.2 Case study from SABRE

Output

The SABRE programme regularly produces realignment ideas that free 10 per cent of currently consumed resources. In medium to large organizations this is quite sufficient to provide a multi-disciplinary team for innovation tasks.

Even in smaller organizations SABRE generates extensive lists of non-vital tasks which can be discontinued, thereby freeing a significant time resource of key individuals.

Concluding comments

Opportunities for business realignment, resource saving and rededication seem to exist in most organizational structures. However, this is perhaps a taboo area. SABRE provides a systematic and yet creative approach to this endeavour and regularly produces free resources which can be reallocated to innovation.

Over a third of all our client companies who move forward into a planned approach to innovation spend nothing further on resourcing their endeavour in terms of indirects. They simply realign individuals already on the payroll to new and innovative tasks. Far from denuding existing business of vital management, we can quote many cases in which the slimming-down process achieved by SPEAR was beneficial to the efficient running of the existing business.

Could it be that some of the systems held as sacrosanct in your organization could be realigned, reduced in terms of resource usage . . . or even discontinued?

PART 4

NEW BUSINESS IDEA GENERATION

SYSTEMATIC
CREATIVITY AND
INTEGRATIVE
MODELLING FOR
INDUSTRY
TECHNOLOGY
AND
RESEARCH

How to convert strategically targeted search areas into specific innovation opportunities.

In Part 4 we present systems, psychology, and a framework for the generation of ideas for new business which are relevant, potentially actionable and capable of generating significant real growth for specific companies in business and industry.

13

Introduction to SCIMITAR

In this chapter we introduce the SCIMITAR system. SCIMITAR is the oldest of the SWORD techniques and has been used by over 700 companies in 200 different industries. It is believed to be the world's most prestigious generator of new business ideas, with over half a million credited to it since 1972.

The need for new business ideas

New business ideas are the seeds of a company's future. Because the life expectancy of any individual business unit is finite, all companies need new business ideas to replace their old products and services at the appropriate time. Strategic planning and creative realignment are necessary but not sufficient steps in the innovation process. The next key step is the generation of specific new business ideas. Moreover, since most companies will face severe competition in the 1990s, the emphasis should be on the word 'new'. There is precious little point in generating an endless sequence of 'me too' ideas for business: this will merely perpetuate competition.

It is very easy to state the need for new business ideas; it is very difficult to create the necessary stream of new business ideas needed in an effective approach to innovation. The difficulty stems from the fact that there will be a conventional wisdom surrounding the current business, which pervades the way the company sees its markets and its technology – its needs and means. The only saving grace is that all the competitors will probably be trapped within the same conventional wisdom. This means that they are unlikely to steal an advantage by a revolutionary breakthrough, but it also means that each generation of new business ideas created within competing companies tends to be very similar, that is, within the confines of conventional wisdom.

What is really needed is to go beyond the confines of conventional wisdom and come up with something that is unconventional, something novel. Novelty will beat conventionality in many competitive situations. If a company can generate a continuing sequence of truly novel business ideas it will seek out, seize and hold field leader position and as such will move beyond the competition. In a word, what is needed is entrepreneurship – see cornerstone 1 in Figure 13.1. In SCIMITAR we seek to copy the mental approaches of successful entrepreneurs.

The numbers game

There is a numbers game at work in industrial innovation which tends to play against individual companies. The often quoted statistic of one successful launch

Figure 13.1   The cornerstones of SCIMITAR

from every hundred initial new business ideas does not seem to us to be too far from the mark. This implies that to find a continuous sequence of launchable new business ideas you need a system within the company which is capable of generating not just a few but literally hundreds of new business options. Few companies have this luxury of choice and many find themselves re-roasting old chestnuts because they simply haven't any new ideas to try.

As we shall see later, systems are available which will reduce the odds involved in industrial innovation, that is, reduce the number of starting ideas required to guarantee launchable successes. Nevertheless, it must be stressed that it is of paramount importance that a company has the ability to generate a relatively large portfolio of initial ideas.

When we ask new client teams how many new business ideas they already have, the average answer is twenty-five. The statistics of the numbers game suggest that with so few to choose from the chances of finding a winner are poor. What is needed therefore is a system that will generate a prodigious number of relevant new business ideas. Consider cornerstone 2 in Figure 13.1 – and start lots of hares running.

Need into means

Beyond the generation of these ideas there is a need for a system which will take them through to the market place, in other words a company must have the means to handle the innovative new ideas that it generates. Hence our use of the word 'integrative': this implies a natural match between the ideas generated and the means of a company. It is pointless to generate new ideas which a company cannot take forward. However, many companies find that they make this mistake. See cornerstone 3 in Figure 13.1.

We therefore differentiate between:

- diversification, and
- integration

Our definition of diversification is more rigorous than that normally used. We use the word diversification to imply the generation of ideas which would take a company outside its current capability in terms of technology and marketing. Conversely, by integration we mean the generation of ideas which can be carried forward to launch within the confines of current technology and marketing capabilities, that is, natural growth points.

Our statistics suggest that the integrative type of new business idea carries a far lower risk than the diversificational type. As we shall see, we can subdivide still further by using a three-dimensional modelling approach to recognize degrees of integration.

Assessment and selection

It is obviously not enough only to generate a portfolio of integrative new business ideas, one must find the winnable concepts. We therefore need a system which is capable, after concept development, of assessing the portfolio. We have found over the years that it is a mistake to try to assess ideas on a once-and-for-all basis. By this we mean that it is far better to assess on a step-by-step basis and equate the severity of assessment with the level of available information.

We present a three-stage assessment and selection procedure for the handling of innovative, integrative new business ideas in Chapter 15. It aims to speed up the process of venture selection and reduce the costs incurred – see cornerstone 4 in Figure 13.1.

Hand in hand with assessment and selection should go the process of concept development. This means that information should be collected to develop the concept in parallel with the appraisal process.

Concept development can take an inordinate amount of resource and time or it can be systematized.

Entrepreneurship

Throughout the process of idea generation, concept development, assessment and selection we perceive the need for an entrepreneurial approach within companies. All too often the multi-departmentalized structure of conventional companies militates against the rapid development, assessment and selection of new business ventures. Entrepreneurial flair, the ability to think your way round problems and to come up with non-conventional solutions, can be the difference between success and failure. Indeed we contend that the private entrepreneurial sector is better at innovation, on a cost-efficiency basis, than the large companies and corporations; hence the Dick Whittington cartoon in Figure 13.1.

We perceive the role of the 'corporate entrepreneur' as crucial to the innovation process. All companies have potential corporate entrepreneurs but many are shunted into backwaters thereby depriving their companies of this vital input. A change in corporate psychology is again required and we shall cover approaches to this in Chapter 14.

Systems and creativity

Our philosophy on new business idea generation, concept development, assessment and selection is that each component part should fit into a sequence of events which is both systematic and at the same time creative. We do not believe that a random approach to these stages in the innovation process can be cost efficient. Certainly if one takes into account the sheer number of new ideas which need to be processed to stand any real chance of generating a continuing sequence of launchable winners, a systematic approach is vital.

In the next chapter we outline the systematic side of the SCIMITAR programme for industrial new business idea generation. In Chapter 15 we outline the creative processes, and in Chapter 16 we present a combined format.

14

Systematic approaches to new business idea generation

In this chapter we outline the systems which have been tried, tested and stood the test of time within the SCIMITAR programme for new business ideas. Readers are referred to Chapter 3 of *Industrial New Product Development* for early applications.

manufacturing. We always refuse! By building one model we produce a unifying structure within which all the departments of a company can see their interrelated roles.

Modelling approaches

In the SCIMITAR system for new business idea generation we use a modelling approach to create a search area for relevant ideas.

As stated earlier, we are strong believers in integrative rather than diversificational ideas. Our case studies consistently show that integrative ideas are more actionable and ultimately more profitable than those which lie beyond the scope of the company. To provide the necessary focus to find integrative ideas we build a model of the company concerned. Why do we use this modelling approach? Models provide a smaller version – a microcosm – of reality, which is easier to study than the complexities of the real company. Models can also focus on specific areas within the system for detailed examination.

We also find that the recommended kind of modelling has another useful attribute; it helps to de-departmentalize companies. We are often asked to build two models for companies, one for marketing and one for

SCIMITAR models

In the SCIMITAR programme we use a three-dimensional form of business modelling. Again, the three dimensions are drawn from the multi-dimensional approach that we use throughout the innovation process. On this occasion the three dimensions chosen are:

- resources
- processes
- markets

We use derivatives of these three primary dimensions to suit the particular company. However, let us first examine the three primary dimensions. These can be used for the most basic types of company which manufacture and market products.

In this simplistic model the products of the company are represented on the base deck of the model as two-dimensional combinations, that is, we define a product as a processed resource. What do companies then do with their products? They launch them vertically and they lodge on individual marketing decks within the model. At this

CAFE 3D

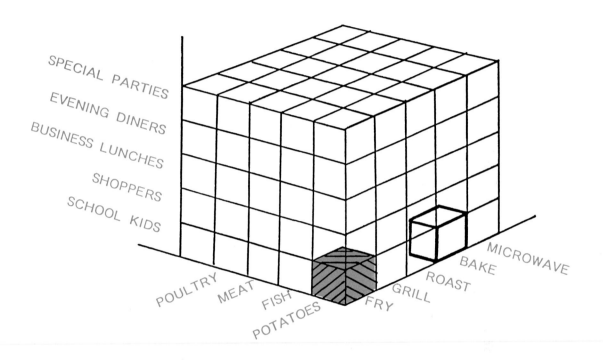

Search for unusual available combinations

Figure 14.1 Basic SCIMITAR modelling

stage the company's business becomes a three-dimensional combination of processed resources sold into specific market niches. In SCIMITAR modelling we frequently build the model in solid form, using perspex sheets and wooden blocks to represent the three-dimensional business units of the company.

Basic modelling

The most basic forms of models are those for very simple businesses: a café is an example. A small café may be regarded as a simple manufacturing business, and we can use the three primary dimensions for its SCIMITAR model, as in Figure 14.1.

Within this model we can define the ongoing business activities of the café. Quite possibly, a small café would make and sell chips – these are represented by the co-ordinate position 1-1-1 in the model. Because this is an ongoing business activity of the café it would be a filled cell in the model.

Further along on the front rank of the model we find another combination, 1-4-1, which represents the combination potatoes, baking and schoolchildren. This may well be void in many café models but it is of significance to a café because it constitutes a three-dimensional definition of a new potential business unit.

This business unit would be typified by preparing baked potatoes and selling them to schoolchildren, say on their way home from school, perhaps as an alternative to chips. As such we may say that combination 1-4-1 constitutes an integrative organic growth opportunity. It is integrative because it lies within the operating periphery of the company; it is an organic growth opportunity because it draws on the skills both in terms of technology (cooking) and marketing (client group servicing).

This simple modelling approach can be used for simple manufacturing and marketing companies. However, for the vast majority of business and industrial concerns it is too simplistic to provide detailed search areas and focusing.

Derivative models

For more complex companies and organizational structures we must use derivative models in the SCIMITAR process. These models are drawn from the three primary variables but use derived axes to add specificity to a company's model. Over the past eighteen years we have tried many different sets of derivative dimensions. By way of examples we can cite the following well-tried and proven derivative models.

- What is being used?
- How is it being made?
- Where is it being used?

These 'WHW' models are derived from:

- Resources – what?
- Processes – how?
- Markets – where?

With a little practice even the most complex of businesses (product- or service-orientated) can be redefined in derivative models. However, it must be stressed that our example of the 'WHW' model is but one of many sets of derivatives we have evolved –it may not work for your specific company.

Searching the model

The derivative models work in exactly the same way as the primary model illustrated in the case of the café. They are also used in the same way: first, one populates the model with filled cells representing ongoing business unit activities, and then searches the remaining voids to find integrative organic growth possibilities.

THREE DIMENSIONAL BUSINESS MODELLING

TRIAL DECK

SCIMITAR Model Deck: X: WHAT / Y: HOW	1 SS M/C	2 Glass Blowing	3 Pipe bending	4 Welded metal	5 Pierced metal	6 Steel working	7 Alum.in.Extru.	8 IR	9 UV	10 Vacuum	11 Direct fired	12 Indir. fired	13 Induct.heat	14 Low pressure	15 Gas fired	16 H₂O₂ fired	17 Fuel oil fire	18 Coal fired	19 DC	20 AC	21 High voltage	22 Vacuum	23 High speed	24 Clean (room)	25 Anaerobic	26 Vapour phase	27 High pressure	28 Steam extract	29 Free radical	30
Pipework 1																														
Containers 2																														
Screens 3																														
Sieves 4																														
Valves 5																														
Structural Work 6																														
Heaters 7																														
Coolers 8																														
Evaporators 9																														
Crystallisers 10																														
Distillation 11																														
Fractionation 12																														
Vaporisation 13																														
Burners 14																														
Calciners 15																														
Deadburners 16																														
Vacuum precipit 17																														
Plasma furnaces 18																														
Vacuum splutter 19																														
Silicon f'n'ces 20																														
Waste Heat Recl 21																														
Filtration 22																														
Microfiltration 23																														
Centrifuging 24																														
Ultrafiltration 25																														
Fermentation 26																														
Bacterial cult. 27																														
Catalysis 28																														
Autolysis 29																														
Ion exchange 30																														
Column reaction 31																														
Fractionation 32																														
Cross linking 33																														
Saturation 34																														
Unsaturation 35																														
Organo metallic 36																														
37																														
38																														
39																														
40																														

Figure 14.2   SCIMITAR search deck

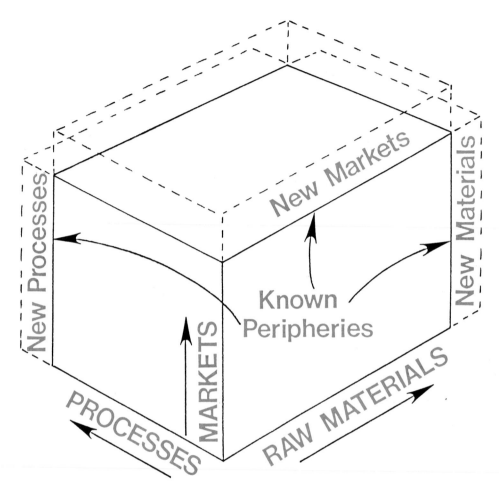

EXTENSIONS HAVE

2 KNOWN PARAMETERS

Figure 14.3   SCIMITAR model extension

By building a perspex model and then downloading this on to paper decks the search is simplified. Existing products are positioned on these search decks (as shown in Figure 14.2) and the voids are then searched.

It will be necessary to persuade the team to postpone judgement and write down any ideas that come out of this systematic search – even if they feel there are good reasons why the ideas won't work.

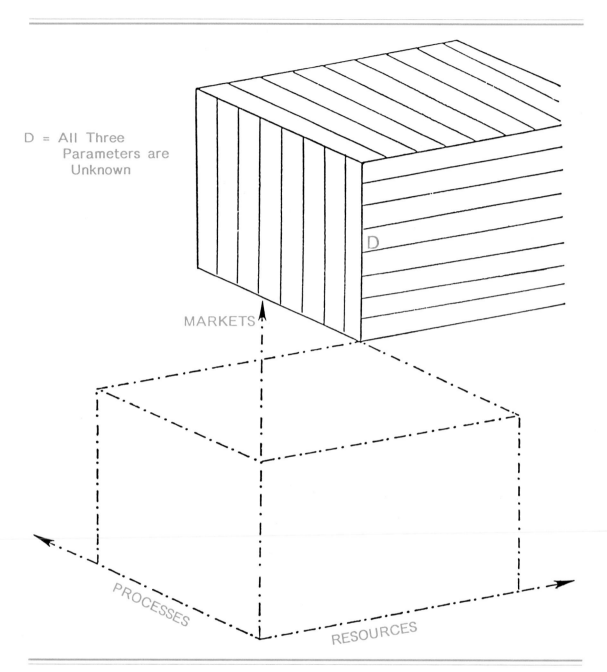

Figure 14.4 The diversificational zone in modelling

Model extension

Once the dimensions of a company's SCIMITAR model have been established they can be extended, and in this way the model can always be kept up to date with the growth of the company. Typically SCIMITAR clients extend their model once a year prior to re-searching it. The developments which have taken place in a company during the year become extensions to the three axes. This is illustrated in Figure 14.3.

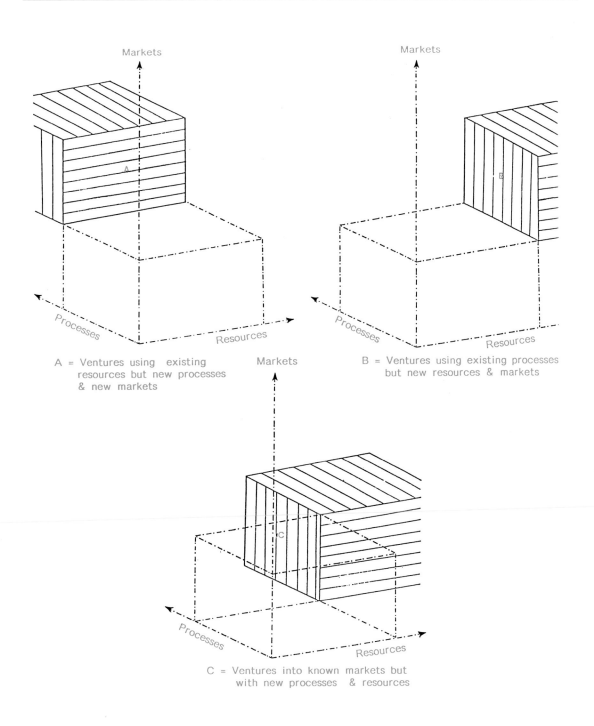

A = Ventures using existing resources but new processes & new markets

B = Ventures using existing processes but new resources & markets

C = Ventures into known markets but with new processes & resources

Figure 14.5   'Darker shades of grey'

These model extensions are like extra slices: each slice has two existing dimensions and one new dimension. As such they are neither integrative nor diversificational but because they have two knowns and only one unknown we suggest that they are still better bets than total diversification. Total diversification implies rushing off and doing something totally new. This is defined in the model by the area perched on the back corner in which all three parameters are new and unknown to the company, as in Figure 14.4.

Would any company be so foolhardy as to rush off and investigate a venture in a situation where it knew nothing about the resource, nothing about the process and nothing about the market? Well, we find numerous examples all the time!

There are of course extensions to the model in which there is only one known parameter and two unknowns. These are represented by the shaded areas in Figure 14.5.

We do not recommend that companies rush off into these areas, as they are nearer to diversification than to integration and we would expect the risks to be higher. By building a model the degree of diversification involved in a specific venture can be defined.

Output

The output from a systematic search of a SCIMITAR model is an initial portfolio of specific new business development ideas of a particular type. They are integrative, that is, they are relevant to and potentially actionable by the client company. As such they advance the cause by taking the strategic search areas defined in SPEAR through to specific, winnable innovation concepts.

Concluding comments

Over the last eighteen years SCIMITAR three-dimensional integrative business modelling has been shown to be a highly efficient approach to the definition of new business ideas.

It must be said that many of the new ideas that emerge from the systematic search of a SCIMITAR model are fairly obvious – once stated! Many ultimately turn out to be 'me toos' because they come from a search area which is within the conventional wisdom of the company and the industry. Nevertheless the systematic search of a SCIMITAR model is a very cost-efficient starting point for the development of an innovation portfolio.

However, it is not sufficient to stop after the systematic search; one must proceed beyond the confines of conventional thinking in an industry if one is to beat the competition with a truly innovative concept.

15

Creative idea generation

Can integrative ideas go beyond the conventional wisdom of an industry? In this chapter we outline the creativity techniques we have built into SCIMITAR to define new business ideas which do go beyond the conventional wisdom of the industry and generate innovation concepts.

Constraints of conventional thinking

One might define a company as a group of like-minded individuals chasing the same agreed business goals. And, while life is never quite that simple, there is definitely a tendency within a company towards convergent thinking. Moreover, since companies must compete in a market place there tends to be a norm within that market place; in other words, there tends to be a conventional wisdom within each industry.

The great difficulty in terms of innovation is that conventional wisdom leads to conventional concepts, which in turn lead to 'me too' products. Such products tend to be commodities, which yield low margins. Conversely, breaking through the confines of conventional wisdom leads to novel rather than conventional concepts. Novel concepts are by definition specialities, which bring higher margins for companies launching them. The company which can launch a continuing sequence of novel products becomes the field leader in terms

of innovation within that industry and can look to a continuing high profit performance. Without this ability companies tend to be run of the mill, and in the 1990s that could spell disaster.

Mind sets

Each company has preconceptions about its situation which, unless challenged, lead to foregone conclusions. The phrase used to define this vicious circle is 'mind sets'; some common examples appear in Figure 15.1.

Unless challenged these preconceptions become the conventional wisdom of the company. Conventional wisdom then gets in the way of the management of change and it becomes impossible to challenge the status quo.

Although we would not argue that expertise in itself is a problem, some experts certainly hinder progress within their companies. An expert is by definition steeped in the conventional wisdom of a specific field and tends to be backward looking, comparing and contrasting all new notions with the established norm.

We believe that to be creative within an organizational structure it is necessary for the ethos of that company to allow its experts to be challenged. Only then can the mind sets of a company be challenged –

 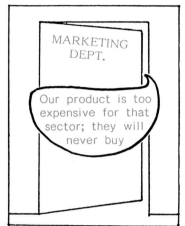 

Figure 15.1   Common mind sets

and only by consistently challenging these can a company keep pace with rapid changes in its sphere.

Over the years we have met many companies in which the experts reign supreme. Far from being characterized by a continuous sequence of expert ideas such companies tend to be followers rather than leaders, and in extreme circumstances are the most conservative laggards of the industry. The rate of change in the 1990s will be so great that such conservatism will prove quite misplaced.

How to break through mind sets

We believe that a powerful way to break mind sets is to bring in an 'amiable amateur'. Some companies are lucky in having an amiable amateur within their corporate structure. These individuals often called the 'joker', are capable of breezing up to the company's experts and asking, politely, 'Why don't you try it this way?' The expert with the mind set usually replies, 'Don't you think I know how to do it? I've been doing it for fifteen years!' In more enlightened companies the amiable amateur succeeds in inducing the expert to try something different. This is a very healthy company ethos! However, it is the exception rather than the rule.

In most companies there is definitely a role for an outsider to act as amiable amateur because the company ethos has militated against such an internal role. This is the role of our SCIMITAR 'lieutenants', that is, my company's operators of the SCIMITAR process. Their role is that of the 'amiable amateur' who knows very little about the intricacies of a company but is capable of seeing round the mind sets and offering alternative suggestions.

The role of consultants in innovation

At this stage it is worth differentiating between the roles of different types of

84

consultants, since many companies turn to consultants for external expertise.

Two types of consultants are active in innovation: 'expert consultants', who have specific know-how relevant to a particular business or industrial sphere, and 'process consultants'. The former can operate only within their sphere of expertise and, since they operate on conventional wisdom, will tend to reinforce mind sets within that sphere. Expert consultants are often disliked because they tend to tell people what they already know, and charge them for the privilege. Process consultants, on the other hand, know little about a particular business or industrial sphere and do not therefore suffer its mind sets. They are experts in their process and can apply it across a wide range of business and industry. In such a way a process consultant can act as the amiable amateur in innovation.

The processes we use in SCIMITAR (and all the other programmes set out in this book) are not specific to any particular business or industrial sector. They derive more from industrial psychology than industrial expertise. They are process consultancy systems.

ears, eyes or nostrils, the human brain has two halves which are not simply mirror images of each other. The left and right hemispheres of the brain are not the same: different mental mechanisms take place in the two halves.

The left brain is the seat of logic; it handles linguistics and numericals and is extremely fast, with a response time of 20-100 milliseconds. The right brain is the seat of creativity, what your grandmother may have called the 'mind's eye', it responds, on average, in between one and two seconds and works by manipulating mental images. It is much slower than the left brain but is reckoned to be many thousand times more powerful.

In terms of problem solving, going beyond conventional wisdom and generating innovative ideas, we should use the right brain. Yet because of the stress of work, which demands instant answers, we rarely create time and space for the slower-moving right brain to generate its creative mental images. We teach teams to recognize the difference between their left brain and their right brain and give them simple exercises so they can experience the difference in terms of mental mechanism and output. Figure 15.2 provides an example.

A framework for understanding mental approaches

Within the SCIMITAR programme the modified psychology or mental approach we use is the well-known hypothesis, 'left brain, right brain'. We recognize that this is somewhat 'old hat' but we are reluctant to change it because it still works exceptionally well in business and industry. So although the hypothesis is now recognized as a gross oversimplification of the complex workings of the human brain, we offer it as a powerful piece of shorthand in explaining the difference between logic and vision.

The hypothesis states that, unlike our two

The expert problem

Once we have explained the different ways in which we can use our brain, we can move towards non-assessive discussion.

Expertise depends on assessment, on the ability to compare and contrast that which is being offered now with that which is in one's memory banks, that is, one's experience. Experts tend to assess against a fixed database, that is, they tend to rely on the left brain. However, most people – even long-standing experts – are able to hold a discussion in a non-assessive mode; to accept statements at face value without

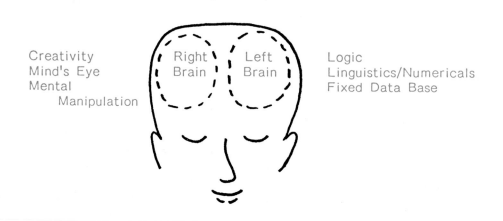

Creativity   Logic
Mind's Eye   Linguistics/Numericals
Mental   Fixed Data Base
       Manipulation

Questions:

1.   What is your address?

2.   Where is the number or name on your house?

Answers:

1.   "I live at ...." the answer to this logical (linguistic or numerical) question comes quickly from the left brain.

2.   "It's nailed to the garden gate" the answer to this unusual question comes from the searching of a mental image in the right brain.

Concept:

The ability to manipulate images in the right brain can take you beyond the conventional thinking of an industry and enable you to conceive novel ideas, e.g. a radio device picked up by van drivers to help find a specific house on a large (un-numbered) estate.

Conventional thinking   - a name is FIXED to the house.

Novel idea   - make the name projectable.

Note   - delivery firms waste £ millions on fuel each year searching for specific addresses.

Figure 15.2   Left brain/right brain test

assessing them and picking out difficulties. Non-assessive discussion is, we believe, the key to creative behaviour: its significance is that one can progress beyond the conventional wisdom of the business or industrial sector. Even half an idea which is generated is well worth writing down notwithstanding that it won't work in its currently stated form. (The great difficulty experts have is the need to tell you all the things wrong with the half idea. However, if trapped and worked on later the half idea might turn out to be either the need or the means in the equation: need + means = opportunity. As such it is well worth recording without assessment as it constitutes a starting point for innovation.)

Again, the problem with experts is that they tend to act as 'devil's advocate'. This may be very powerful in some situations but it is quite destructive in creative ones. It is not necessary at that early stage to criticize an idea, it is far more desirable to add to the idea even though you may believe fundamentally that the basis of the idea is wrong. Many a wrong idea has led to an industrial breakthrough.

Non-assessive discussion

Through explaining the workings of the left and the right brains to teams we have found that they can indeed suspend judgement for the duration of a creativity exercise within the SCIMITAR programme, and prevent experts from trampling all over partly formed ideas. Once they grasp the importance of non-assessive discussion company experts become vitally important to creativity exercise. Once, that is, they start to contribute in a non-assessive way. Here the role of the amiable amateur is vital in that one can challenge overstated expertise thereby opening up the discussion and going beyond the conventional wisdom of the industry.

Once the guidelines for a non-assessive discussion have been established we can introduce specific creativity techniques to engender a powerful forum for idea generation.

Creativity techniques

Over the years we have tried out over fifty different creativity and creative problem-solving techniques within the SCIMITAR format. Some techniques seem more powerful than others in breaking mind sets. We have found that many of these techniques have a common approach: the provision of mental triggers that stimulate non-conventional images in the right brain – the mind's eye. These can then be manipulated through non-assessive discussion to create ideas that go beyond the mind sets of the situation.

We have gradually refined our techniques to the point where we have a standard structure. (No doubt this 'standard' structure will evolve further!) It uses different techniques from many different sources, drawing from the following well-known creative problem-solving techniques.:

- Attribute Listing
- Brain Writing
- Lateral Thinking
- Synectic Excursions

Specifically, the SCIMITAR programme uses brain writing as the main vehicle and superimposes on it attribute listing, lateral thinking and synectic excursions. In Figure 15.3 we present a typical SCIMITAR creativity exercise.

It will be seen that the brain-writing sheet is a format within which a strange mental trigger can be tested. Where do we find these odd mental triggers? We have produced many lists of mental triggers generated through attribute listing, lateral thinking and synectic excursions. In Figure 15.4 we provide a typical list of mental

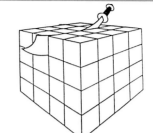

NATURALISTS CLUB ACTIVITY

SYNECTICS

PROFITEERS TEAM

INSTRUCTIONS: Fill in the next box and pass it on

1. Let's focus on Deck: ___Building Products___

    In particular, let's consider (Product Zone); *Cement based*

    If, in this zone, we could (like Nature): *Make leaves change colour with Seasons*

2. This might mean:

    Building cladding that changes colour with time of day or year

3. A possible line is:

    Different coloured "leaves" sprayed on

4. An alternative line is:

    Different coloured "tiny tiles" hung on hooks

5. Why don't we:

    Small plastic leaves in stucco

6. Why don't we:

7. I can picture the new product as:

    A coloured spray-on stucco to change the colour of buildings

8. This could be improved by:

    Making the surface dirt shedding

9. I think we could make and sell:

    A range of tinted cement paints to improve the colour i.e. visual appeal, of high rise concrete structures in cities.

DON'T ASSESS

Figure 15.3   SCIMITAR creativity exercise sheet

MARKETEERS' TRIGGERS

If only we could offer:

Thinking products
Products that see ahead
Self-repairing products
Mating products
Products that reject competitors
Products that grow
Products that die quickly
Infectious products
Lovable products
Furry products
Products with bells
Child's play products

TECHNOCRATS' TRIGGERS

What happens if we make:

Cheap and nasty products
Rejects
Mixed up products
Products that fall apart
Talking products
Products that run backwards
Hellishly expensive products
Products that don't work
A universal product
Foreign style products
Antique products
Idiot products

PROFITEERS' TRIGGERS

What if like nature we sell:

Products that change with seasons
Sexy products
Products that hibernate
Products that give birth
Products that eat one another
Diving duck products
Hunting products
Chameleon products
Snail products
Kangaroo products
Mother-in-law products
Tasmanian Devil products

Figure 15.4   SCIMITAR mental triggers

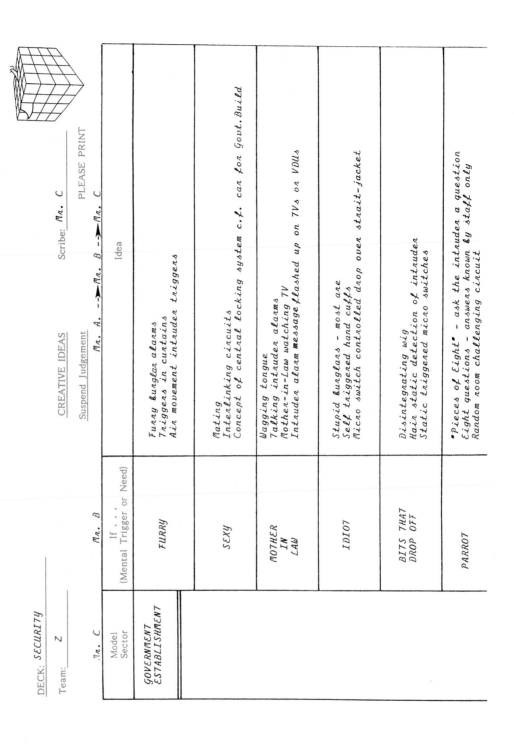

Figure 15.5   SCIMITAR creative output sheets

triggers which we use with SCIMITAR client teams.

By using these and other lists of mental triggers, we are able to go beyond the conventional wisdom of a business or industry and get people to see things in a different light. Out of this SCIMITAR creativity exercise regularly come 'world beaters' – concepts which are bigger than the total combined business of the client company.

Focus

There is a skill in selecting mental triggers for the SCIMITAR creativity exercise. The triggers must refocus the team's attention on possibilities relevant to their company, otherwise the 'world beater' will not be actionable. We achieve this by using the SCIMITAR model to generate triggers. Resources, processes and market definitions are taken from the axes of the model and then disguised as triggers. In this way a list of powerful metaphors is produced which triggers creative and relevant mental images within the team.

By way of an example, Figure 15.5 shows a set of disguised needs drawn from their model by a SCIMITAR client team. These are the 'If . . .' statements. These triggers are then worked on in three-person teams. Each produces a sentence to develop the concept and . . . a real idea emerges.

Concluding comments

By using this approach to focused creativity within SCIMITAR we are able to generate many new ideas that go beyond the conventional wisdom of the client company, yet they are integrative. We have also found that we can generate ideas of quality at a remarkably good speed. Our target is one fundamentally new idea per team member per minute during the creativity exercises. In this way we regularly create 500 new ideas beyond the conventional wisdom of a client company – in one day during the SCIMITAR programme.

However, in the final analysis, it is not the systems which are important in creativity but the attitude of mind. We can provide a framework for understanding the workings of the human brain but we cannot force anyone to be creative. We do still meet 'abominable no-men' – individuals who are totally resistant to change in any form. Nevertheless we have found that the role of the 'amiable amateur' can challenge many mind sets and improve the creative output of most teams.

The creativity technique outlined above has now been used by more than 3000 managers in business and industry; it has stood the test of time. When used on day 3, the creativity day, in our SCIMITAR programme, it never fails to produce hundreds of new but relevant ideas.

16

A programme for new business idea generation

Both SCIMITAR modelling and the creativity techniques outlined can be used individually to generate new business ideas. They can also be used as an interlocking programme, which we have developed for idea generation, concept development and venture selection. The SCIMITAR programme is outlined in this chapter.

Format

We have repeatedly faced two questions when dealing with the SCIMITAR programme: 'Surely we can do it ourselves!' is the first. However, many clients came back to us having tried to do it themselves. We can say only that the role of the 'external amiable amateur' seems to be vital. The other question is: 'What will we do with all the ideas generated?' Companies feel the need for a combined programme – first half, ideas generation; second half, idea assessment and selection – that produces a short-list rather than just an initial portfolio.

Our SCIMITAR format is a six-day programme set over three months. The first three days generate a multitude of relevant new business ideas using modelling and creativity – we are sufficiently confident to guarantee a minimum of 250 new business ideas, irrespective of client or business sector! The second three days provide a multi-stage concept development, assess-

ment and selection procedure which seeks out the ten or so most viable ventures from within the initial portfolio.

Although the SCIMITAR programme is flexible, we have learnt some hard lessons as to the sequences of events that should be followed and we set out our guidelines in the following sections. The format we use is presented in Figure 16.1. The modelling system and creativity exercises we use in this programme are as set out in Chapters 14 and 15. These exercises comprise the first three days of SCIMITAR.

Separating idea generation from idea assessment

Time and again we have seen teams fall into the pitfall of failing to separate idea generation and idea assessment: they slide out of non-assessive discussion into 'devil's advocacy' without realizing it. This effectively torpedoes ideas at the moment when they surface. At that moment of creation they are certainly not strong enough to fight off a devil's advocate. At best they are only half ideas and usually ill-defined at that. We therefore insist that there is an interval between the end of the creativity exercises in SCIMITAR and the commencement of concept development, assessment and selection.

It is a mistake to go straight from idea

SCIMITAR SIX DAY PROGRAMME

AGENDA

Day 1	Session 1	Introduction to SCIMITAR System
	Session 2	Introduction to Integrative Modelling
	Session 3	Company Modelling
	Session 4	Review of Data

One month intervention . . . Data Collection

Day 2	Session 1	Review of Data
	Session 2	Modification to Model
	Session 3	Preparation for Systematic Search
	Session 4	Systematic Search

Day 3	Session 1	Introduction to Creative Thinking
	Session 2	Introduction to Creativity Techniques
	Session 3	Creativity Techniques
	Session 4	Creative Search

One month intervention . . . Data Collection

Day 4	Session 1	Approaches to Assessment
	Session 2	Initial Selection
	Session 3	Concept Development
	Session 4	Data Review

One month intervention . . . Data Collection

Day 5	Session 1	Introduction to Second Stage Selection
	Session 2	Second Stage Selection
	Session 3	Introduction to Third Stage Selection
	Session 4	Third Stage Selection

Day 6	Session 1	Introduction to Venture Justification
	Session 2	Development of Venture Justifications
	Session 3	Presentation of Ventures
	Session 4	Overview of the SCIMITAR Process

Figure 16.1   SCIMITAR programme format

generation into rigorous assessment. We insert an intermediate stage. This has the effect of building on the idea to the point where it can be more fairly judged.

Half ideas

Many of the ideas generated in the idea generation exercises within SCIMITAR are only half ideas. They are either a 'pious hope' for example a need in a market place which cannot easily be fulfilled, or a 'discovery push', that is, means that are available without a natural usage or need. It is therefore necessary to balance the equation:

- Needs + Means = Opportunity

This requires concept development to bring the two half ideas together.

Some idea interpretation is also often required. Ideas as written down in either systematic form or creative form can be misleading. We cite an example in Figure 16.2.

What is required is a concept development exercise that can piece together these ill-defined ideas so as to balance the opportunity equation.

Concept development system

The concept development system that we use in SCIMITAR is based on 'Mind Mapping'. This is an exercise in which all the facets of an idea can be drawn together at one time. A typical mind map is shown in Figure 16.3.

The SCIMITAR Team of an electrical instrumentation company were amused to discover the following void in their model:

Resource	=	Solar Energy
Processes	=	Illumination
Market	=	Coal Mines

They were persuaded to record this non-logical combination:

"Solar powered miners' lamp!"

The following creative re-alignment took place subsequently in the SCIMITAR Programme:

"Use of light energy generated by miners' cap lamp to illuminate a small 'solar panel' on a fixed non-powered instrument. The instrument will work when observed but be passive otherwise - no switch, no spark, no explosion hazard."

Figure 16.2   The solar-powered miners' lamp

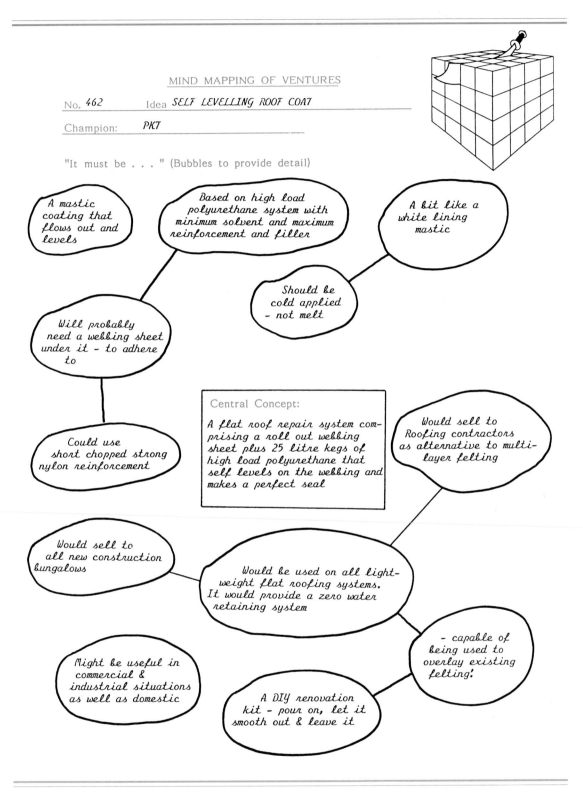

MIND MAPPING OF VENTURES

No. *462*   Idea *SELF LEVELLING ROOF COAT*

Champion: *PKT*

"It must be . . . " (Bubbles to provide detail)

A mastic coating that flows out and levels

Based on high load polyurethane system with minimum solvent and maximum reinforcement and filler

A bit like a white lining mastic

Should be cold applied - not melt

Will probably need a webbing sheet under it - to adhere to

Could use short chopped strong nylon reinforcement

Central Concept:

A flat roof repair system comprising a roll out webbing sheet plus 25 litre kegs of high load polyurethane that self levels on the webbing and makes a perfect seal

Would sell to Roofing contractors as alternative to multi-layer felting

Would sell to all new construction bungalows

Would be used on all light-weight flat roofing systems. It would provide a zero water retaining system

Might be useful in commercial & industrial situations as well as domestic

A DIY renovation kit - pour on, let it smooth out & leave it

- capable of being used to overlay existing felting!

Figure 16.3   Mind mapping

On this mind map you can see the one-liner as produced in the creativity exercise at the top of the page. Around the central stage – the box in the middle – are all the facets of the ideas drawn together from a multi-disciplinary team and set out as variations on the theme. At a later stage of mind mapping the essence of the idea is distilled out of the mind bubbles and written out in the central box.

We liken mind mapping to 'theatre in the round', where all the actors can be seen around the stage at the same time. In this way anyone in the audience can see what is happening on the stage and can take a balanced view. This balancing of the view, or balancing the equation, is crucial to concept development prior to assessment and selection.

Principles of assessment and selection

There are many pitfalls which must be avoided in the assessment and selection of ideas for business innovation. For example, we believe that the systems used should balance the severity of assessment with the level of available information. It is all too easy to assess too harshly at too early a stage. Many poorly defined but actually quite good ideas can be lost in this way.

The cardinal sin, to us, is to assess on a once-and-for-all basis. Yet many companies do exactly this in terms of their budgeting procedure for new business ventures. They assess once a year and a venture either receives a budget or does not. The idea may be ill-defined at budgeting time, in which case it will be discarded, will not be budgeted for and therefore cannot be worked on to improve the definition. Dare we call research and development budgeting 'annual myopia'?

To avoid these pitfalls we define a multi-stage assessment and selection procedure

within the SCIMITAR programme. Typically this has three stages.

First stage assessment

SCIMITAR generates an average of 600 new business ideas for client companies. All the ideas that come out of both the systematic and creative search are recorded, refined as a statement and then subjected to this first stage appraisal. This first stage appraisal takes the form of three questions which are asked of each of the ideas in the initial portfolio:

- Is the idea new to the company?
- Is the idea relevant to a market need?
- Is the idea actionable by the company given its budgetary constraints?

We retain in the portfolio only those ideas which are new, relevant and actionable, because we are looking for new business opportunities which are integrative organic growth options. The statistics of the assessment procedure are shown in Figure 16.4.

Second stage appraisal

The second stage appraisal used in SCIMITAR is a form of risk analysis. (Between the first stage and the second stage it is necessary to collect further information to underpin the concepts still further. This information can be conveniently logged on the mind maps. As we have said, we believe it is necessary to equate the severity of assessment with the level of available information.)

Conventional Risk Analysis in industry tends to be uni-dimensional, that is, financial risk analysis. We prefer two-dimensional risk analysis, taking into account:

- *Procurement risk* – the risk involved in making or buying in the product (or service).

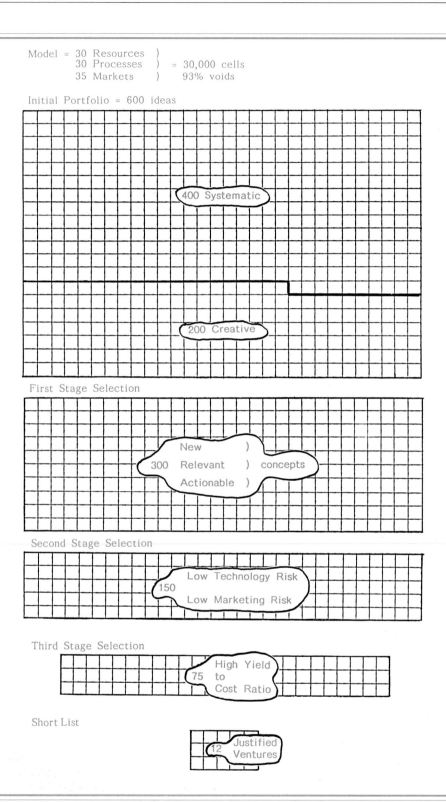

Model = 30 Resources)
       30 Processes) = 30,000 cells
       35 Markets) 93% voids

Initial Portfolio = 600 ideas

400 Systematic

200 Creative

First Stage Selection

300 New)
    Relevant) concepts
    Actionable)

Second Stage Selection

150 Low Technology Risk
    Low Marketing Risk

Third Stage Selection

75 High Yield
   to
   Cost Ratio

Short List

12 Justified Ventures

Figure 16.4 Typical SCIMITAR statistics

```
TECHNOLOGY

    1.      Could make it now on existing plant

    2.      Minor doubts that we could modify plant
            to make it

    3.      Major doubts that we could ever make
            it i.e. new technology


MARKETING

    1.      Could sell it now without difficulty

    2.      Minor doubts about selling it with existing
            sales force

    3.      Major doubts that we could ever sell
            it e.g. conflicting market
```

Figure 16.5   Risk analysis

● *Marketing risk* – the risk involved in marketing and selling the product (or service).

We have evolved a simplistic approach to risk analysis which involves setting up risk scores. By examining previous ventures, both successful and failed, it is possible to establish risk scores that reflect those things a specific company finds difficult and even frightening. In Figure 16.5 are some typical risk scores from SCIMITAR exercises. Each venture is scored 1, 2 or 3 on each axis and the scores are then multiplied together to give a one to nine risk rating. Low score equals low risk, and these are the ones selected for further appraisal.

Third stage appraisal

Again, there should be an interval between second and third stage appraisals so that further information can be collected. In this way we equate the severity of the assessment with the level of available information. The information that is collected can be conveniently stored on the mind maps.

In the third stage of appraisal we estimate the yield that will be created by the venture in the first three years after launch, and set this against an estimate of the costs that will be incurred in taking the venture through to launch. Yield and cost appraisals at early stages in ventures are notoriously inaccurate:

● 'A company's initial estimate of the potential of a new market turns out, on average, to be 300 per cent of reality.'
● 'A company's initial estimate of the costs that will be involved in taking the

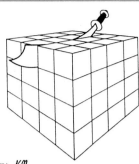

V.A.S.T.

No. *649*   Idea: *One handed can opener*   Champion: *KM*

Marketeers: POTENTIAL YIELD   (Additional Sales)

1. Market Size   *UK and Western Europe (alone)*   = 10^6 units pa
2. Ruling Price – *current price of competitors' two handed*   = £ *2.45* /unit
3. Total Market Value   ($\frac{1 \times 2}{1000}$)   *product*   = £*2450* K pa
4. Targeted Market Penetration (after 3 years) *Should get*   = *10* %
5. Sales expected in 1st year   *1%*   = *10K*   = £ *24.5*K pa
6. Sales expected in 2nd year   *5%*   = *50K*   = £ *122.5*K pa
7. Sales expected in 3rd year   *10%*   = *100K*   = £ *245* K pa
8. TOTAL Sales in 3 years (5+6+7)   [(8a)= *160K*   = £ *392*   K

Technocrats: POTENTIAL COST (Physical Spend)

9.   D & D Costs   *1.0*   man years @ £ *25*   K pa   = £   *25*   K
10. Pre Production   *0.5*   Man years @ £ *25*   K pa   = £   *12.5*   K
11. Launch (liaise with Marketeers) – *give it some oomph!*   = £ *100*   K
12. Cost of New Plant (if needed) *existing No. 2 Plant*   = £ *--*   K
13. Cost of Mods. to existing Plant (if needed) *6 new moulds*   = £ *18*   K
14. Additional People Costs (   ) = £ *--*   K
15. Other "one off" costs (*£? Probably not*) = £ *--*   K
16. TOTAL "Attributable" Costs to Venture (9+10+11+12+13+14+15) = £ *156*   K

Profiteers: LIKELY CONTRIBUTIONS

17. Likely Direct Cost of Units *Assembly £0.15   Metal Parts £0.25*   = £ *0.60* /unit
      Plastic £0.20
18. Likely Gross Margin (2–17)   = £ *1.85* /unit
19. Likely Contribution over 3 years (18x 8a) *£1.85 x 160K*   = £ *296*   K
20. Yield/Cost Ratio (19÷16)   *£296K ÷ £156K*   = *1.9*

Approx. Pay Back Period (Ratio of 1·0 = 3 years) *3 ÷ 1.9*   = *1.6* years

---

Figure 16.6   VAST appraisal

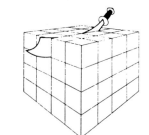

VENTURE JUSTIFICATION REPORT

VENTURE NO: 426	IDEA: *Lorry type Blow-out alert*	CHAMPION: *VJS*

1.	THE OPPORTUNITY	*To develop a niche market for established technology in a core business area.*
2.	THE PRODUCT	*A tyre wall temperature T/C linked to cab buzzer*
3.	RESOURCES/**COMPONENTS**	*Standard Type 2 T/Cs plus telemetry*
4.	THE PROCESS	*Assembly only plus packaging*
5.	THE MARKET	*Heavy goods vehicles & fleet operators*
6.	MARKET SHARE → SALES	*Year 1 10,000 units @ £40)* *Year 2 25,000 units @ £40) £2.4M sales* *Year 3 25,000 units @ £40) in 3 years*
7.	COSTS OF VENTURE	*R & D = £25K Assembly line parts = £60K* *Specification = £25K Other Costs = £10K* *TOTAL £120K*
8.	LIKELY CONTRIBUTION	*£25 x 60,000* *Margin expected to be £25 per unit therefore yield = £1.5M in 3 years.*
9.	THE UNKNOWNS	*1. Market acceptance* *2. Exact positioning & design*
10.	DEVELOPMENT NEEDED	*1. Detailed Market Survey -- APRI* *2. R & D plus D & D*
11.	CONCLUSIONS (SUBJECT TO ANSWERING THE UNKNOWNS)	*A low risk viable venture which is integrative with Technology and Market*
12.	RECOMMENDATIONS	*Investigate*

Figure 16.7   VJR format

new venture through to launch turn out, on average, to be 50 per cent of reality.'

How then can we be accurate in terms of yield and cost?

Our answer is that we cannot be accurate at such an early stage in concept development. However, we believe we have solved the conundrum. What we have found is that one can substitute consistency for accuracy. By appraising all 100-plus surviving ventures in the space of a day the human brain has to fall back on its optimum mode of operation, namely comparing and contrasting. The human brain is not very good at assessing absolute values in isolation but given a pack of 100 ideas it is quite good at comparing and contrasting within that pack.

We would not therefore vouch for the absolute accuracy of our third stage appraisals but we would vouch for the consistency of the order of priority produced. The format we use, 'Venture Appraisal and Selection Technique' (VAST) is shown in Figure 16.6.

We normally find that there are at least seventy-five ventures above the VAST acceptance line: a luxury of choice! Ultimately we prefer to aim for a balanced portfolio of about twelve ventures from well within the company's pay-back guideline at the end of a SCIMITAR three-stage appraisal and selection procedure. Within these twelve we prefer four rapid low-cost, low-risk ventures for immediate action; four medium-term growth ventures; and four longer-term main growth points within which we hope there will be one 'world beater' – a venture which at maturity will be bigger than the company itself.

In a typical SCIMITAR exercise we would expect the team to generate 600 ideas in the initial portfolio. The three stages of appraisal and selection which follow the concept development exercise of mind mapping tend to yield a very wide choice, usually between fifty and a hundred potentially viable ventures. This luxury of choice provides for a very rigorous final appraisal. The ventures which are ultimately submitted for consideration in the final short-list are:

- new to the company
- relevant to a market need that can be serviced by the company
- actionable by the company given its budgetary constraints
- tolerably low procurement risk
- tolerably low marketing risk
- capable of giving a high yield to cost ratio
- a shorter pay-back period than the company guideline

Indeed, on the last point, the great majority of ventures that come through the SCIMITAR procedure imply a pay-back inside six months as against the company criterion of three years. So, if ultimately our estimate is 300 per cent out in terms of yield and 50 per cent out in terms of cost, that is, it is 600 per cent of reality (which we doubt), it is still in line with the three-year pay-back guideline (multiply 6 months' pay-back by 6, equals 36 months' pay-back, which is the norm!).

The final twelve or so ventures are then written up at the end of the SCIMITAR programme, in a convenient form known as a Venture Justification Report, a typical example of which is set out in Figure 16.7.

The SCIMITAR programme has been proven in various guises over the past eighteen years. During that time we have worked with over 800 companies on a worldwide basis, in over 200 different industries. SCIMITAR has generated over half a million new busines ideas, each one, by definition, of relevance to the company concerned. We believe that this proves the validity of a systematic yet creative approach.

The final output from SCIMITAR is a short-list of highly interesting, justified and actionable ventures. These then go forward to the next stage in our systematic approach to innovation, venture development and launch – RAPIER.

Finally there is an added bonus from SCIMITAR. We train your in-house team so that they can re-run the programme on an annual basis without any further input from the consultants. In this way a continuum of new business ideas can be created.

Concluding comments

If your company is short of ideas that it can develop organically into important new ventures, you need a system which is capable of generating hundreds of relevant ideas and then further systems to help you pick the winners.

Most companies have only a couple of dozen really new ideas to choose from and no system for finding more. Most companies never have sufficient ideas at any one time to prove the validity (or otherwise) of their assessment procedures.

PART 5

EXTERNAL SOURCED GROWTH POINTS

BUSINESS
   LICENSING AND
      ACQUISITION
         DEVELOPMENT
            EXERCISE

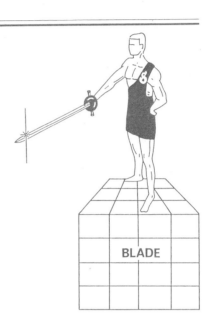

BLADE

There are of course many companies in both business and industry for whom internally sourced new business ideas would not be actionable because they simply do not have the resources to carry such ideas through to fruition – even if they used SABRE. In Part 5 of the manual we present an alternative to organic growth: an approach to the location of new business opportunities outside the company.

17

An introduction to external searching for growth opportunities

Organic growth, as outlined in Part 4, is only one source of ideas for business innovation. Ideas do not need to come from within a company, they can be sourced externally; however, this requires different techniques. The licensing or acquisition of tranches of business are examples of such techniques.

Different sources of business innovations

There is a fundamental difference between organic growth on the one hand and acquisition and licensing on the other; the latter are externally sourced growth points whereas the former involves sourcing ideas internally.

For many companies organic growth is the obvious way forward. Ideas can be generated, developed and launched from within the company and thus the next generation of business units is internally sourced. If the internal sourcing option – organic growth – is available to your company, you should stick with it. Better to stay with what you know than run the risk of bringing something in from the outside which could turn out to be alien to your corporate culture.

However, for some companies organic growth is simply not viable: they may not have the wherewithal to develop ideas

generated inside the company, or they may be so far removed from the leading edge of relevant technologies that to build up the necessary development capabilities would be far too costly. In these cases it is sensible to look to externally sourced growth points.

As the company does not itself generate the ideas it follows that different techniques are needed to find them. External ideas can be found from either acquisition studies or licensing studies, or a combination of both.

Definitions

Our definition of acquisition is the buying in of new business units by buying in an established company entity. Licensing, we define as the licensing in of new business units, or the right to operate those business units. Both acquisition and licensing are therefore similar in that they are externally sourced growth points.

Difficulties

Difficulties, even disasters, are legion in the field of acquisitions and licensing. It may well be true to say that very few acquisitions work *per se*, the vast majority are cost efficient only after a fairly rigorous rationalization process to cut away the bad

105

which was brought in with the good. Such 'nights of the long knives' need to be avoided on a cost-efficiency basis if no other. However, we believe that the transient damage to company morale and corporate structure goes far deeper. Therefore we argue that one should study a potential acquisition long and hard before going ahead, to check that it really is complementary rather than alien to your corporate structure.

The same applies to licensing in a new product or service line. It seems to be true that few companies will license out their 'cash cows' until they have been well and thoroughly milked! This may mean that you are licensing in a concept which has run its natural life and which has very few further profit-earning years to come. These are to be avoided.

The conglomeration approach

During the 1970s and 1980s many large business groups were brought together as a conglomerate, that is by acquiring a wide range of dissimilar business units.

While many of these were successful we believe that the scope for such a conglomeration process is now limited. Many of the companies acquired by these conglomerates were 'black dog companies' and to improve their efficiency it was necessary only to slim them down to produce an improved return on capital employed. Large companies could do that only for as long as black dog companies existed. As we move into the 1990s we expect these opportunities will gradually dry up. Moreover, it can be argued that this process is by definition inefficient, in that it requires wasteful rationalization. It is also dehumanizing for those who are acquired and then discarded.

We argue that it is far more efficient to look for a truly synergistic approach to

acquisition and licensing, in which natural linking points are sought out between that which is being bought and your existing corporate strategy and culture.

Synergistic approaches to external sourcing

We believe that externally sourced innovation options can be either integrative or diversificational, just as can organic growth.

The word synergy is often used in connection with externally sourced ideas: 'Is there any synergy between what you are buying in and your core business?' We equate the word synergy with integration; in other words we believe that the right kind of externally sourced idea is integrative rather than diversificational. The question is how to detect whether an externally sourced idea has synergy or not. Our answer is that you cannot appraise it on a once-and-for-all basis. You need to collect information gradually about the idea and appraise it several times so as to equate the severity of the assessment with the level of available information. We do not believe that rushing into either a licence or an acquisition is likely to be low risk. Better, we think, to embark on a less hurried 'Courtship Approach'.

We will present our 'courtship approach' in the next chapter. It may seem like a slower approach to external sourcing of ideas, which we accept, but our experiences show that it is far less risky than rushing straight into a 'quick marriage'.

The BLADE approach to acquisition and licensing

Just as rushed marriages don't always work and it is better to get to know your prospective partner before proposing, we

1. It can help to avoid wrong marriages, i.e. the acquisition of companies which simply do not fit the corporate entity, because they are not synergistic or, stated another way, they are diversification acquisitions rather than integrative.

2. Although the process may seem long, it can give rise to a rapid initial impact through licensing options prior to acquisition. This can give a finite impact to bottom line profitability very quickly.

3. It allows companies to contemplate major ultimate goals, i.e. acquisition of major companies in a step-wise and non-threatening way. This means that small but rapidly growing companies can contemplate major acquisitions at some stage in the future.

4. The BLADE approach is less disruptive of an organisational structure than is conventional acquisition strategy. By the time the acquisition is actually taken, the incoming structure is well understood and can be smoothly integrated into the corporate whole.

5. Each stage in the process of BLADE can be profitable in its own right and the long term profit impact of BLADE can be at least as large as conventional acquisition approaches. Indeed, we believe that on a cost efficiency basis, BLADE wins.

6. Because BLADE is a sequential approach, companies can gear the level of involvement to the level of available resources thereby avoiding over-stretching themselves. In this way, even small companies can play.

7. If the courtship doesn't work out at any stage, for example at the licensing stage, then you can stop short of full acquisition. In this way, it becomes a self selecting approach.

Figure 17.1   Advantages of the courtship approach

believe it is better to get to know a potential acquisition through an intermediate licensing stage rather than rushing straight out and buying. As with organic growth, where we believe that companies should generate a luxury of choice before deciding which ventures to embark on, we believe that licensing and acquisition should be chosen from a wide-ranging group of options. Most companies rarely have more than four acquisition options to choose from at any one time. We believe this figure should be forty, or even 400.

As we shall see, it is possible to use licensing as an intermediate stage so as to check out the validity of a potential acquisition; in other words, companies should court their potential partners before rushing to acquire them.

The BLADE approach has three stages:
1 Licensing = 'first date'.
2 Joint ventures = 'engagement'.
3 Acquisition or merger = 'marriage'.

The BLADE approach generates a large range of potential licensing options, leading to a smaller range of potential joint ventures, which in turn leads to an even smaller range of possibilities in terms of acquisition and/or merger. Only if the first two stages work well would companies need to contemplate the third.

Potential advantages

The courtship approach has a range of potential advantages over more conventional approaches to external sourcing. These are set out in Figure 17.1.

However, to realize these advantages you need to generate a luxury of choice in the early stages. As we shall see BLADE is capable of multiplying the number of options available to most companies in terms of externally sourced growth points.

Integrative external sourcing

At first sight 'integrative external sourcing' may seem a contradiction in terms: 'How can anything which is externally sourced be integrative?' We believe that there are many integrative options available to companies in terms of external sourcing.

The most obvious example of an integrative external source is the concept of acquiring a competitor. Indeed, during the 1970s and 1980s the vast majority of conglomeration-style acquisitions included the acquisition of competitors. This is fine in many respects except that it can be rather limiting, as a specific company has only a limited number of perceived competitors. It is also limiting in another respect: acquiring one's competitors may not be possible because they may be the wrong size, poor or unavailable, you may run into difficulties with cartel or monopoly commissions, or there may be market barriers to prevent the conglomeration of one's competitors. Finally, the acquisition of competitors may give rise to management differences and difficulties – not least because of the antipathy born out of competition.

What then are the alternatives to acquiring one's competitors? The BLADE approach suggests alternative integrative external sources. Companies may, for example, contemplate the (eventual) acquisition of companies servicing their sector or companies which are different from themselves but complementary. Once this concept is grasped the constraints associated with acquisition short-lists disappear. Any given company will have many complementary potential acquisitions in associated fields.

We therefore suggest that in addition to looking towards the acquisition of competitors one should also contemplate the acquisition of 'complementaries'. The advantages of acquiring complementaries include technology transfer, the capture of

new and related markets, and bringing in new management blood in related fields. The ultimate justification for the acquisition of complementaries rather than competitors is that this approach to acquisition generates real growth within an industrial sector, unlike the conglomeration approach.

We liken the acquisition of complementaries to marrying a girl in the next village whereas the analogy for acquiring one's competitors would be marrying one's first cousin.

The BLADE techniques

The BLADE programme is a step-by-step courtship approach to externally sourced innovation options. In the next two chapters we examine the modelling and creativity approaches used, and in Chapter 20 we outline the format for a fully interlocking BLADE programme.

18

Systematic approaches to external opportunity sourcing

In this chapter we set out the models and other systematic approaches we use in the BLADE programme.

Having covered that ground, the next form of modelling is unique to BLADE. We will concentrate on this new form of modelling in this chapter.

Modelling approaches

In BLADE we use two separate modelling approaches, one after the other. We strongly recommend that companies which are about to embark on acquisition and licensing – seeking externally sourced opportunities – should first use the SCIMITAR model previously presented.

The SCIMITAR model is advantageous even when sourcing externally because it creates a far more detailed search area and understanding of a company's internal skills and capabilities, which is of vital importance when going out to court synergistic, that is integrative options. Readers should therefore study the three-dimensional model set out in Chapter 14. This model should be constructed first, using the three primary dimensions – resources, processes and markets – and then by deriving specific dimensions from them, such as 'What?', 'How?' and 'Where?' (particularly useful derivatives in this context). Using the SCIMITAR model will give rise to a host of further options for growth and development which can be sourced either organically (internally) or externally through acquisition and licensing.

BLADE modelling

Again, BLADE modelling is drawn from our basic theory of the multi-dimensionality of business, and readers are referred to Part 1 of the book. For BLADE, the three dimensions we choose from the multi-dimensional array are:

- products
- markets
- sources

Products are self-explanatory in manufacturing terms but need to be explained if we are considering service sectors. Most service companies can envisage their packages as 'products'; indeed many refer to them as such and we have rarely found difficulty in this definition in BLADE modelling. However, if it helps a derivative such as 'service packages', 'processed resources' or even 'what is sold' can be used.

Note that this axis is an amalgam of two primary dimensions, sources and processes.

'Markets' are self-explanatory though again you may have to use derivatives such as

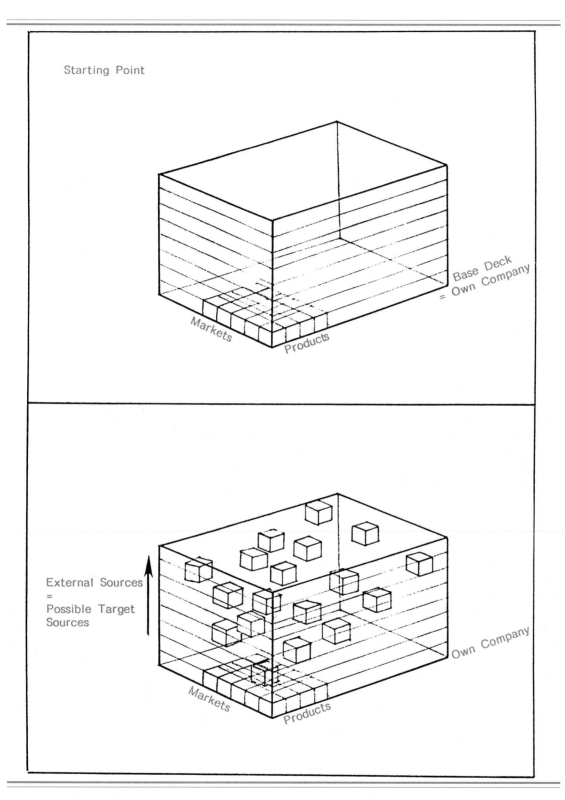

Figure 18.1   BLADE modelling

'client groups' or 'where'. By 'sources' we mean external sources – other companies from which the products can be obtained. They can be obtained in the interim period under licence and subsequently by outright acquisition.

The approach to modelling is to collect as much information as possible about all products, markets and sources of relevance to the client company. Here it is important to avoid mind sets and to recognize that one can build into the model complementary companies as well as competitors. This is one of the reasons why we recommend that a SCIMITAR model is built as a precursor to the BLADE model. This initial model will have the effect of breaking down the mind sets and allowing the team to perceive all sorts of complementary technologies and markets.

Once all the relevant products, markets and sources have been listed they are set out along the three dimensions of the BLADE model. Some care is required in choosing the definitions and in working up derivative definitions both for the axis and for individual items along it. Only in this way will the model really come to life when it is searched.

A typical BLADE model will create a three-dimensional matrix containing about 20000 cells. This implies many thousands of triple combinations of potential business options. Again, only a very small percentage of these will be filled. Where are these filled cells? It has become a convention that in terms of BLADE modelling the base deck of the model represents the company itself. So at the start of the exercise the filled cells are all on the base deck of the model and represent the company's specific products and their sale into specific markets. Above the base will be decks representing other potential sources: potential acquisitions or at the very least companies from which products could be licensed. Figure 18.1 shows the model.

It will be seen from Figure 18.1 that the model actually grows outwards from a restricted base. This is because the overlaying decks in the model extend both product options and marketing options. In this way the BLADE model depicts the core of the current business and provides search areas for ever expanding options.

Searching the model

Once the model has been built and extended to a sensible horizon, stopping short of companies which cannot realistically be envisaged as external sources, the team's task is to search it.

The first stage is a purely systematic affair. Each of the decks representing specific companies is searched in turn. This is conveniently achieved by laying that particular deck immediately above the company's own deck to allow for 'visual projection' of the existing company activities on the base deck upwards on to the overlaying deck. Visual projection asks the question: 'Is there any linkage between our current activities and the company under consideration?' If the potential source is 'integrative', synergy will immediately be apparent from the model since the existing product range will be complemented by the product range of the source. Moreover, the markets serviced will tend to complement existing markets. In this way the degree of synergy is maximized. For convenience, we use paper decks to study individual potential source companies. On these decks we 'project' filled squares to represent the client company's activities and 'X' to indicate source company activity. This purely systematic search generates intriguing 'overlapping patterns' as exemplified in Figure 18.2.

This systematic search generates a short-list of potential acquisition options. However, you do not have to acquire the targeted source company to acquire business options. There are many less binding ways to engender business.

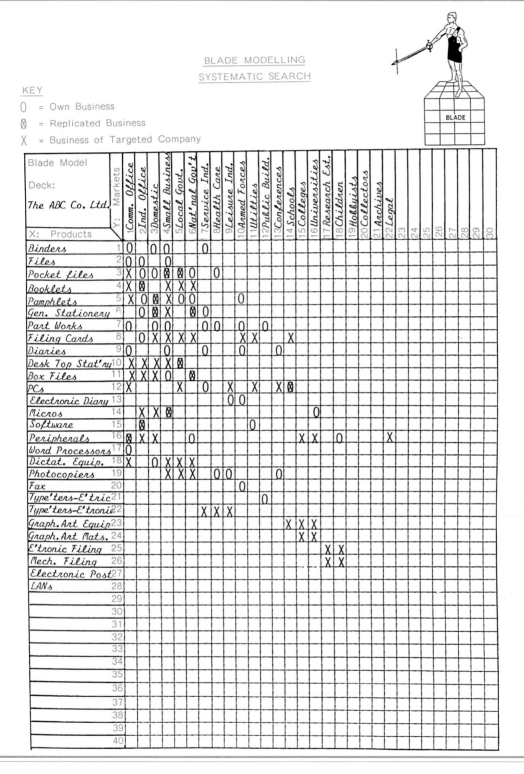

Figure 18.2   BLADE model systematic searching

Concept interpretation

At the end of the systematic search of the BLADE model an initial portfolio of options will have been generated. At this stage these will be no more than three-dimensional

merchanting

trading

buying in

sub-contract manufacture

LICENSING

franchising

exclusive licensing

back to back trading

JOINT VENTURES

capital participation

management exchange

dual launching

equity participation

ACQUISITION/MERGER

Figure 18.3   Approaches in courtship

word combinations: a product for a market from a source. As such they will require further interpretation. It may be sensible to revert to the SCIMITAR model to crack open the product definitions and redefine these in terms of technological approaches and product attributes, thereby providing the necessary interpretation.

Concept interpretation is vital at this stage because shortly we must start to appraise the concepts.

During the concept interpretation procedure various sub-options will arise, representing various ways of engendering new business. In Figure 18.3 we provide examples. These various stages in the courtship approach lead from one to another in a programmable, manageable sequence. At each stage there is the opportunity to redefine the sub-option so as to maximize its cost efficiency at that time during the BLADE approach.

Certainly you don't need to rush into marriage if you use the courtship approach!

Further steps

The BLADE programme is concerned with the first few steps in the external innovation staircase. The higher ones will be covered later in this book. However, before developing the external business options further it is important to go beyond a purely systematic search and find those intriguing potential partners 'in the next village'.

In the next chapter we present a creativity technique to aid the search.

19

Creative approaches to external opportunity sourcing

'Can you find some more creative licensing/ acquisition opportunities, preferably ones our competitors have missed?' In this chapter we present a creative technique for opening up the search area for relevant externally sourced growth points which neither purely systematic nor conventional approaches will discover.

Conventional licensing and acquisition thinking

There is a problem with conventional licensing and acquisition thought: it tends to focus on very similar companies to one's own, that is, it tends to suggest the acquisition of competitors rather than complementary companies.

The difficulty stems from the fact that most managers charged with responsibility for licensing and acquisition studies within companies, have spent considerable time not only in their own company but within a specific industry. In this way their thinking processes will tend to be introverted and 'mind setted'. We have worked with many hundreds of company executives charged with responsibility for externally sourced growth, and have noted that as time goes by they tend to become more blinkered! They tend to become experts in licensing and acquisition searching within their industry but, as we have said previously, expertise can sometimes get in the way of creativity.

Again therefore we see the need for the amiable amateur who is prepared to challenge the perceived wisdom of a situation as to which sources are relevant.

In the 1970s and 1980s many companies could not decide whether vertical or horizontal integration was better. One of the spin-off benefits of the BLADE model is that it provides a three-dimensional search area within which both horizontal/vertical and lateral integration options are perceivable. This in itself can open up the search area. However, we do not believe that a purely systematic search of the BLADE model is itself sufficient. There will be large 'no go' areas within the decks of the model because of the mind sets of the company executives searching it. Typical mind sets we have heard expressed are presented in Figure 19.1.

These and many other mind sets need to be challenged, politely, if options which have eluded the conventional wisdom of your industry are to become part of your BLADE portfolio.

Creative approaches

Again, it is first necessary to provide a framework for understanding the nature of mind sets. This is achieved in BLADE by providing teams with either the 'front of mind/back of mind' or the 'left brain/right

115

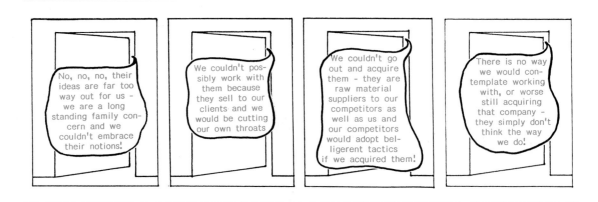

Figure 19.1   Typical mind sets in external sourcing

brain' hypothesis, for which readers are referred to the appropriate sections in Chapters 7 and 15.

Once a suitable hypothesis is in place it is possible to work through a range of creativity techniques in an open-minded non-assessive way so as to prise open the 'no go' areas within the BLADE model.

A confidence-building point at this stage is to reinforce the list of forms of business available via the 'courtship approach' (see Figure 18.3). At this stage nobody is suggesting that you should go into partnership with a particular company. There are many stages during which the synergy between you and the potential source can be checked before anything even as firm as a licensing agreement needs to be formulated. Therefore you can with confidence check a whole range of potential sources without being too fearful of the ultimate scenario.

Having tried out many different creativity techniques within BLADE programmes we have settled on one in particular, and we present this below.

Four-Dimensional Brain Writing

We have so far approached the search for innovation on a multi-dimensional basis using a selection of three dimensions at a time to define the search area. In creative external searches for innovative opportunities we believe it is necessary to embrace four dimensions. It seems to us that an externally sourced innovation opportunity comprises elements drawn from:

- market – need
- product – means
- source – company
- approach – form of business

These four dimensions are drawn from the six which constitute the business complexity. However it is again necessary to use derivatives of the primary divisions so as to make the search specific.

Having built and searched a three-dimensional BLADE model based on market, product and source, and having taken into account the various forms of business available along the 'courtship approach', it is possible to stretch the imagination to provide and search a four-dimensional model. Only by thus stretching the imagination have we found really intriguing options which go beyond the conventional wisdom of a specific company/industry. However, to achieve this it is necessary to coach teams in four-dimensional thinking.

CLOVER LEAF EXERCISE on: *Substrate Specifics* No: *24*

Actions:

1. Marketeers – set the target
2. Technocrats – define products for the target
3. Profiteers – select a possible source company
4. Team – define First Courtship Approach

1. MARKETEERS

To complete our spectrum of outlets, we should penetrate the newly emerging DIY/ BIY sector for substrate specific adhesives.

In particular, we should target their use in masking i.e. as a substitute for tapes.

2. TECHNOCRATS

To capture business in the DIY/BIY market, we need to make our product more "user friendly".

An appropriate approach would be the presentation of our range in "squeeze gun" applicators.

3. PROFITEERS

There are only three established companies who fill and sell "squeeze gun" style special mastics and glues: X, Y & Z Cos.

Interestingly, Y Co. Ltd. are very active in the masking tape market.

4. TEAM

Consider a "back to back" approach in which we trade in their masking tapes – to learn the market and then contract fill our S.S. adhesives in their "squeeze guns".

CONCEPT: *Enter the DIY Mastic Sealant Market with a "squeeze gun" applied range produced and own branded for use by Y Co. Ltd.*

Figure 19.2   The four leaf clover game

Find the four leaf clover

The technique we use to coach teams in four-dimensional thinking is a simple paper exercise using the device of a four leaf clover design, as shown in Figure 19.2.

As we have said at several points in this book, we have found that in trying to alter the standard psychology found in business and industry so as to produce a more creative output, the most powerful techniques seem to be among the most simple. Indeed we wonder whether modern industrialized society could learn much from the simplistic learning systems used in primitive cultures and by children in our own society.

In searching for a simplistic technique to accompany the 'courtship approach' to external sourcing of innovation opportunities, we wondered whether the games played by young lovers could be relevant. In nineteenth-century England lovers spent many hours searching for four leaf clovers. In that most clovers have three leaves the discovery of a four leafed variety was believed to be very special and indicative of a special relationship between the finders.

Spatially a four leafed clover is very useful in our quest for creative ideas. It allows teams to consider the juxtaposition of four dimensions all represented on the same sheet of paper. Teams are invited to test ranges of combinations of these four dimensions (drawn from the BLADE model and the 'courtship approach') to discover intriguing combinations that allude to novel externally sourced innovation opportunities.

Playing the game

The Four Leaf Clover Game is very simple to play. The first step is to define the search area by using the model to focus on a specific market need which is within the orbit of the client company. In this exercise the marketeers in the team take the lead. They name a clover leaf by defining a specific need in the market place that they can service – given an externally sourced product.

The next step is to pass the clover leaf to the technocrats, who define a specific product which would satisfy the need. At this stage the technocrats are trying to link known technology to a market need. This is not as difficult as it may sound in that although the company will not have the product required to satisfy the need the technocrats within the team will be aware of the technology required.

The third step is to pass the clover leaf to the profiteers, whose job is to list potential sources – companies who could possibly be approached to supply products of the need/means definition defined in the lobes of the clover leaf. There is a link between the technology required and the potential sources and the profiteers will need to liaise with the technocrats during this activity.

Once this third lobe is completed the whole team concentrates on the fourth and final lobe, picking a form of business which is appropriate. Remembering that forms of business can be set out sequentially, as in the 'courtship approach', it is not difficult for a team to progress along this sequence and choose the most advanced but acceptable form of business.

Output

The output from the four leaf clover game is often amazing. Teams frequently discover external opportunities for innovation which have always been there but which have never been focused and therefore never really been recognized. Once recognized they constitute a truly integrative opportunity for innovative new business of a low-cost, low-risk nature.

Typically BLADE teams generate hundreds of four leafed clovers. The knack is to write down as many potential opportunities as possible by suspending judgement at this stage. At a later stage the really viable options can be selected. However, if teams slide into judgement while playing the four leafed clover game their creativity will be inhibited. The options produced will require some interpretation as they will tend to be rather nebulous at this stage.

20

A format for external sourcing of growth opportunities

The BLADE model, the sequence of the 'courtship approach' and the four leaf clover game can all be used as free-standing management systems, or they can be used within a composite format. In this chapter we present the composite format that we use in the BLADE programme for generating and assessing a portfolio of external options for client companies.

Format

The BLADE programme is a six-month campaign targeted to produce a short-list of at least ten externally sourced new business opportunities. The sequence of events followed is as shown in Figure 20.1.

The approach used to modelling and creativity is as set out in Chapters 18 and 19. After these steps come the assessment and selection procedures, outlined below.

Approach to assessment and selection

We contend that the approach to assessment should be first to develop the concept as much as possible and only then to assess it. Therefore the first part of the BLADE assessment and selection procedure is concept development.

Whereas in internally sourced organic growth options there may be difficulty in finding information to underpin concepts, there is no such difficulty in the BLADE programme. All the necessary information is available because the business units already exist; the only difficulty is that they exist in another company. What is needed is to prise out the necessary information about other companies. This may sound a little like industrial espionage but it is totally legal!

Our approach is to collect as much information as possible at this first stage. This allows us to be fairly caustic in our assessment because we have collected plenty of information on which to base it. Given the luxury of choice that BLADE produces, teams can be rigorous in assessing the output from their initial portfolio.

Data sourcing

Where do we find the information with which to underpin the external source concepts? It exists in company publications in the form of annual reports and glossy brochures on individual product ranges. We collate the information in a three-dimensional format derived from the BLADE model. For further useful sources readers are referred to Appendix II in *Industrial New Product Development*.

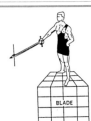

BLADE SIX MONTH PROGRAMME

AGENDA

Day 1, Month 1 Session 1   Introduction to BLADE System

Session 2   Introduction to Integrative Modelling

Session 3   Practical Modelling

Session 4   Data Targeting

Day 2, Month 2 Session 1   Review of Data

Session 2   Modification to Model

Session 3   Preparation for Systematic Search

Session 4   Systematic Search

Day 3, Month 3 Session 1   Introduction to Creativity Techniques

Session 2   Creativity Techniques

Session 3   Creative Search - "Four Leaf Clover" game

Session 4   Introduction to Assessment System

Day 4, Month 4 Session 1   Approaches to Assessment

Session 2   Initial Selection

Session 3   Courtship Development

Session 4   Data review

Day 5, Month 5 Session 1   Introduction to Second Stage Selection

Session 2   Second Stage Selection

Session 3   Introduction to Third Stage Selection

Session 4   Third Stage Selection

Day 6, Month 6 Session 1   Courtship Planning

Session 2   Venture Justification

Session 3   Presentation of Ventures

Session 4   Overview of the BLADE Process

Figure 20.1   The BLADE programme

Appraisal

Once the necessary data is collected appraisal can take place. The appraisal system used in BLADE is virtually identical to that used in SCIMITAR, for which readers are referred to Chapter 16. The approach has three stages:

1 *Initial appraisal*
   Is the concept new, relevant and actionable (given the proposed form of business)?

2 *Intermediate appraisal*
   Is the concept tolerably low risk
   ● to procure?
   ● to market?

3 *Final appraisal*
   ● What yield will accrue from the venture?
   ● What costs will the venture incur?

These three stages will reduce the initial portfolio progressively and provide a final short-list of potentially viable options.

Statistics and case studies

The BLADE six-month programme typically creates two models: a SCIMITAR model of 30000 cells and a BLADE model of approximately the same size. Searching these two models gives rise to an initial portfolio averaging 600 external innovation options. The statistics of assessment and selection are set out in Figure 20.2.

These 600 options are developed through the information-collecting procedure set out in this chapter, which has the effect of weeding out some non-starters. Typically about 200 interesting options are fed into the three stages of assessment during BLADE. Each stage tends to reduce the portfolio by 50 per cent, thus yielding approximately fifty potentially viable

ventures to choose from at the end of the exercise.

Because this is too many for the average company to proceed with, teams are usually asked to make a final selection on the basis of their preferences and expertise, and this gives rise to a final short-list of ten ventures for presentation to the company's board on the afternoon of the final day of the BLADE exercise.

Some of the most poignant case studies we have collected through the use of BLADE revolve around the breaking of mind sets. A common mind set in manufacturing industry is the belief that companies must make everything that they sell. Once a BLADE team recognizes that they can find and capture externally sourced opportunities which obviate production difficulties, the sky *is* the limit. An example, disguised for confidentiality reasons, is presented in Figure 20.3.

This breakthrough of a common mind set has been dubbed the 'non-involvement factor' within BLADE programmes. Recognizing that you do not have to get involved in making a product is like a breath of fresh air to a manufacturing company. Faced with serious competition difficulties many large companies have benefited greatly from this non-involvement factor.

Concluding comments

The external sourcing of innovation opportunities can be regarded as either a very simple or a very complex approach. Its simplest form is the acquisition of a competitor. However, we have argued that this simplest form is not necessarily the most effective form of external growth. By using modelling approaches and superimposing four-dimensional creativity it is possible to widen the search area for externally sourced opportunities and focus

Average output from BLADE Programme

Initial Portfolio

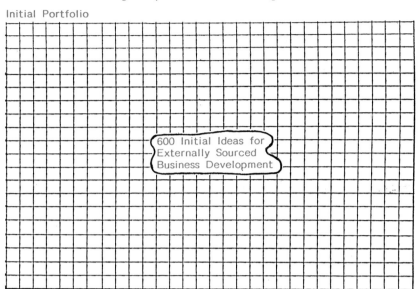

600 Initial Ideas for
Externally Sourced
Business Development

Developed Concepts - Interesting Options

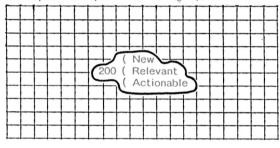

200 (New
(Relevant
(Actionable

Second Stage Appraisal

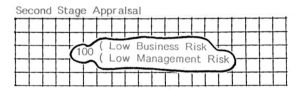

100 (Low Business Risk
(Low Management Risk

Third Stage Appraisal

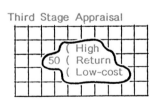

50 (High
(Return
(Low-cost

Output:

10 justifiable externally sourced ventures

Figure 20.2   BLADE statistics

A relatively small Company in the Snack Industry carried out a BLADE exercise in 1986. Out of this came the recognition that new technology was about to sweep their industry. Several leading competitors already had the technology - which carried a price tag too great for the Company concerned.

However the "Four Leaf Clover Game" turned up a potentially interesting external source for new business. This involved the courting of a French Company (who had the technology).

When approached, this external source were delighted by the possibility of gaining a foothold in the UK and readily agreed to produce a range for the Company concerned.

This range was produced (using the advanced technology) and bulk packed abroad, then down packed on existing equipment into the final pack in the UK. The range was launched - and prospered - as did an ever broadening range of back to back deals between the two companies.

Perhaps this will lead to a union between the two companies in 1992 . . .

Figure 20.3   A BLADE example

on options of a far more dynamic and creative form than is usually the case in acquisition-orientated growth strategies.

If your company needs to acquire to grow you need a systematic and creative approach to find the right company to acquire, and we recommend you court them cautiously rather than marrying them quickly!

PART 6

VENTURE DEVELOPMENT

RAPID
  ASSISTANCE
    PROGRAMME FOR
      IMPLEMENTATION OF
        EXPERIMENTATION AND
          RESEARCH

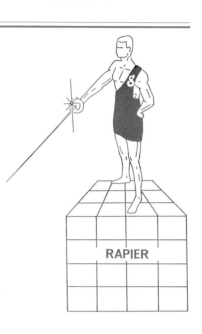

RAPIER

'How do we develop these innovative growth concepts?'

Whether new business ideas are discovered internally or externally the next stage is developing those concepts through to launch. Part 6 of the manual presents management systems for taking ideas through development to the point where a decision can be made to launch them.

21

Approaches to venture development

In this chapter we outline our approaches to the complex and difficult task of venture development. For most companies venture development is harder than idea generation.

Introduction

Experience has shown that for most companies idea development is far more difficult than idea generation. Many companies are perfectly adequate at generating a continuing stream of new venture ideas but fall down time and time again on their development. However, we would argue that these difficulties may be caused by choosing the wrong ideas in the first place. Analysis of failed ventures regularly indicates that highly diversificational ideas had been chosen. Again we wish to stress our differentiation between integrative and diversificational ideas: the former are far easier to develop than the latter.

The often quoted statistic of one successful launch out of every hundred ideas seems to us to be about right in unstructured approaches to venture development. Faced with such poor odds most companies would give up, especially if the spend on the ninety-nine failures is high. It is worth investigating why venture development is so notoriously difficult. We certainly feel this is a vital endeavour and have spent much of the last ten years developing systems. For an outline of the origins to our approach readers are referred to Parts 2 and 3 of *Industrial New Product Development*.

Difficulties

Our case studies suggest that there are many and varied difficulties associated with venture development. However, three recurring themes, shown in Figure 21.1, are worth passing on as cautionary tales at this stage.

When one examines individual case studies a common denominator emerges. Many ventures fail through a breakdown in the communication systems within the company developing the venture.

Conventional approaches

As we will see later, we are fundamentally opposed to the conventional approaches used by most companies for venture development. First, however, we should set down what we have found to be the convention.

Over the past ten years we have had the opportunity to study venture development approaches used in over 400 companies in over 100 different industries. Although, as one might expect, there is a great variety in

---

```
1. · Misunderstanding the Market

     Many ventures fail because the item which is
     developed does not match the market need.


2.   Falling Short on Technology

     Developing a product with a specification signifi-
     cantly below that which is required by the market
     place.


3.   Getting Resourcing or Timing Wrong

     Under-resourcing ventures so that they starve
     or stall.
```

Figure 21.1   Common difficulties in venture development

individual approaches, it is possible to set down 'the conventional approach'. In a phrase the conventional approach is 'Sequential Venture Development'. By this we mean systems which pass ventures through a sequence of departmental activities.

The research phase

Typically venture development starts in research and development departments. They work on the fundamentals of the venture for perhaps six months to a year, in extreme circumstances anything up to ten years. During this time the venture is solely their preserve. Indeed in many situations there is a positive attempt to prevent the communication of ongoing venture information until such time as the research and development phase is completed. It is no laughing matter to hear remarks such as: 'don't tell the other departments about it until we know it will work.'

At the end of the research and development phase the idea has reached what we call 'product in a bottle'. At that stage it is written up in report form.

The development phase

The venture then finds its way to the next department in the chain. Although there has been a gradual change over the last decade, most companies can still be characterized as 'production-led' (as against 'market-led'). In production-led companies the second department in the chain is production. It receives the venture in report form from research and development and sets about it. Unfortunately, comments such as 'Look what these idiots from R & D want us to make now' are not uncommon.

Production works on the venture for six months to a year – in extreme circumstances up to five years. The aim seems on many occasions to be to prove that the 'product in a bottle' cannot be made. More enlightened departments will struggle gamefully to make the product in a bottle in isolation from research department, which has effectively closed the book on the venture and probably cannot help anyway because the venture no longer has a research budget. (Remember our name for budgeting – 'annual myopia'.) For a small number of ventures at the end of the production/works exercise a proven pathway has been found to a manufactured product; for the majority, however, the venture stalls at this stage.

The marketing phase

The surviving ventures are then passed to the third department in the chain, usually marketing, which picks up the venture, again usually in report form backed by some semi-production scale samples. It in turn sets about the venture, possibly with the cry: 'Look at this rubbish that production have made – how do they expect us to sell this?'

Because the product has been developed in total isolation from the market place it is indeed quite likely to be rubbish, in the sense that it cannot be sold for the original purpose envisaged. Under such circumstances it is perhaps not surprising that marketing departments tend to file it under 'forgotten'. More enlightened departments may make a genuine attempt to force-fit the embryonic product into the market place. This is a very difficult task because the exact requirements in the market place have long been forgotten, misinterpreted or even deliberately ignored. Many ventures fail at this stage. For the lucky survivors with a tenuous link between the product as defined and the market need as initially conceived there is almost certainly some mismatch which will require further work in production. However, by now production's feet are set in concrete. It knows how to make the product, it has proved it can make the product and it has written a report to prove it – it is also very reluctant to change its approach. This can kill off the venture.

The financial phase

Finance is the next and usually final department in the chain. If the venture has survived this far, it is placed on the desk of the financial controllers. Financial resources must now be found to gear up both production and marketing for the venture launch. Because finance has not been involved at any previous stage (except in terms of budgeting) it will probably find that the venture does not lie comfortably with the company's corporate strategy. In addition the resources required may be out of line with resource availability.

One final area of difficulty in multi-divisional companies is to find that the venture treads on the toes of another division.

Conventional output equals 99 per cent failures

This may appear a very jaundiced view of the conventional company approach to venture development – it is! We do find some more enlightened approaches, usually among those companies which are market-led rather than production-led. However, even here they use a similar sequential approach, the only difference being that

CONVENTIONAL APPROACH to Project Development	RAPIER APPROACH to Project Development
R & D Phase typically 6 months to 1 year = £30K	Concurrent Development using a Multi-Disciplinary Team

Piloting & Trials
    typically 3 to 6 months   = £15K
	Time to launch
	    – 6 month Programme

Test Marketing
    typically 6 to 9 months   = £20K
	Typical development staff costs
	    12 people for 26 working days each
	                                    = £32K

Trial Production typically 6 to 9 months = £20K	

Launch
    Time to launch
            minimum 21 months
            maximum 36 months
	Time to launch = 0.5 years

Typical development staff costs = £85K	Success rate over the period
	    1978-1981 = 16 successful launches
	    out of 36 ventures drawn from
	    1240 ideas (ex SCIMITAR)

Time to launch   = 2.5 years
 	
Success rate over the period 1972-1977
    = 2 successful launches out of
    31 ventures drawn from
    104 ideas	

Example cited is taken from a company's new product range extension exercises in the retail horticultural sector.

Figure 21.2   RAPIER comparison

marketing has the first shot, followed by research and development, production and finance in that order. So there is still plenty of scope for disasters.

The point we are making is that sequential venture development is fundamentally wrong.

Alternative approach

Faced with this débâcle of what has gone before in many of our client companies we have gradually evolved an alternative approach. It suggests that venture development should be a concurrent multi-disciplinary approach rather than a sequential chain of individual disciplines.

By this we mean that all the disciplines of the company should be involved at the start, throughout venture development and be together at the point of launch. This alternative approach is intended to remove interfacial barriers and to prevent communication breakdown during venture development. It has the additional spin-off of improving team spirit and co-operation within the company, which transcends venture development and permeates the whole of the company's ethos.

We are fundamentally opposed to the idea that a venture can be the sole preserve of any department at any stage during its development. We try to point up the folly of interfacial barriers leading to departmental rivalry which slows down, disjoints or kills off ventures during development. And we stress that the risks involved in venture development, which in any event are high, can be reduced by this multi-disciplinary concurrent approach.

We can prove that our approach works successfully because in RAPIER we see one successful launch out of every twelve initial ventures. We can also prove its cost efficiency because the spend on the eleven ventures which will ultimately be shelved is far less than it would otherwise have been. A comparison between conventional approaches and RAPIER is shown in Figure 21.2.

Overview

This alternative approach to venture development has been named RAPIER. As we will see, RAPIER is an interlocking sequence of management systems which take innovative concepts through venture development to a point where they can be launched. The various management aids within RAPIER can be used either as free-standing individual options or as a composite format.

Of all of the alternative management systems we present in this manual, RAPIER is the most significantly different approach when compared to the current norm in business and industry. Conversely, one can say that in terms of overview it is venture development that we feel is the most inefficient sector of the innovation process at the present time. Whereas other components of our integrative innovation procedure can make worthwhile savings when compared to the existing approaches, RAPIER can make truly massive savings.

In the next chapter we present the modelling approach that we use in RAPIER so as to provide a systematic framework for venture development. In Chapter 23 we then provide an alternative psychology to venture development and in Chapter 24 we present a composite format for the use of all the RAPIER techniques.

22

Systematic approaches to venture development

This chapter contains a modelling approach to venture development. The RAPIER model presents a systematic framework for a multi-disciplinary concurrent approach.

Approach to modelling

RAPIER models are constructed from three dimensions drawn from the six-dimensional approach to business set out in Part 1. For venture development exercises we choose the following three modelling parameters:

- venture – concept
- resources – RAPIER team people/time resources
- quests – developmental questions to be answered

These axes are complex derivatives of the six basic dimensions of business. Traditionally we build the RAPIER model with the venture dimension vertically so that each venture has its own deck.

Data collection

Once the RAPIER team is established most of the data required for the RAPIER model are available. RAPIER teams are normally extensions of the teams used for SCIMITAR or BLADE.

It is important, however, to feed in all potential ventures at this first stage, as it is difficult to add them at later stages in the proceedings. The RAPIER team's first activity, in terms of modelling, is to compile an exhaustive list of those ventures which are to be handled in the exercise.

Because, on average, only one launchable venture will be found within each twelve, we always like to start with thirty-six to forty-eight. Therefore all ventures which came out of SCIMITAR and/or BLADE should be listed. It may be that there are other ventures available to RAPIER, for example those which have stalled in previously operated systems.

It is far better to start the RAPIER exercise with fifty ventures than ten. Given fifty it is reasonably certain that five will ultimately be launched and that by the end of the six-month programme the first will be very close to launch. However, it is necessary to stress that this does depend on the gestation period of the industry. (We run year-long programmes in some industries.)

Team commitment

The second axis in the RAPIER model is team resource. The standard commitment expected of individual members in RAPIER teams is one day a week for the duration of the six-month programme. They are thus committing themselves to approximately

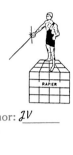

QUEST MAP

No: ___10___ Idea: *Pigmentary Oxides* _____ Discipline: *Technology* Author: *JV*

I don't know if we can capture the *right* RM for the *right* green

We know of one source but it's only 250 tpa and it might be *variable*

What about testing or colour matching?

If the RM varies, we'll have a hell of a job colour matching

Sequence of Quests
1. Define target product
2. Re-appraise RMs
3. Trial Calcinations
4. Colour examination
5. Discussion of Options

Where do XCC Ltd. get their starting material from?

They make it from ore!!

What product are the Marketeers asking us for??

If we can't locate that Irish source we'd better think in terms of smaller tonnage and higher price

Could we make a W grade equivalent?

Figure 22.1 Quest maps for RAPIER

twenty-six days each (double this on the year-long programme). The average RAPIER team numbers twelve and therefore the total man day availability to the RAPIER programme is approximately 300 in six months.

For convenience we usually divide the team resource axis into either twenty-six or fifty-two slots – one per week. Within each slot there is 'a model within the model' to represent individuals' activities. (Each primary slot in the axis therefore represents twelve man days.)

Development actions

Along the third dimension of the RAPIER model we set out all the quests necessary to take concepts through to launch. This quest dimension is divisible into three groupings:

- actions to be taken by marketeers
- actions to be taken by technocrats
- actions to be taken by profiteers

However, we do not try too hard to differentiate between these three activity groups as the aim is for a multi-disciplinary approach.

Although we have evolved some guidelines as to the types of actions necessary for each discipline in taking ventures through to launch, it is important to stress that no two company teams give rise to the same list of quests. The differences stem from the differences in approach within the company coupled to differences in requirements within different industries. We therefore present the team with pro-forma sheets and ask them to devise their own step-by-step 'quest plans'.

Quest plans

Quest plans are like Mind Maps (see Figure

22.1). We use them to collect as much interrelated information as possible on the various steps required in the venture development exercise. Each discipline produces its own quest plans and we then amalgamate them to produce a composite listing of actions required, which we then set out along the third dimension of the RAPIER model.

Establishing the model

Once all three axes have been defined the model can be established. Again, we build the model out of perspex sheets to give a three-dimensional entity. Each perspex sheet represents a specific venture. However, the model can be read horizontally, vertically and laterally. By this we mean you can take slices through the model in any of the three planes. As we shall see later, this is useful in maintaining the three-dimensional approach. Once the model has been established in three-dimensional form it can be converted to paper decks. In this way we generate 'chequer' sheets where two of the three dimensions are represented on the X and Y axes. A typical deck from a RAPIER model is shown in Figure 22.2.

Venture chequer sheets

By representing a specific deck in the model as a chequer sheet we define all the actions and available team resource for that venture. As various actions are then carried out against the venture, its progress can be plotted on the sheet. The purpose of the chequer sheet is to make sure that we do not omit any actions which might subsequently prove to be important. Only in this way can we be sure that all the necessary actions in support of a venture have in fact been carried out.

Not all actions will be appropriate to all ventures. Nevertheless they are all plotted

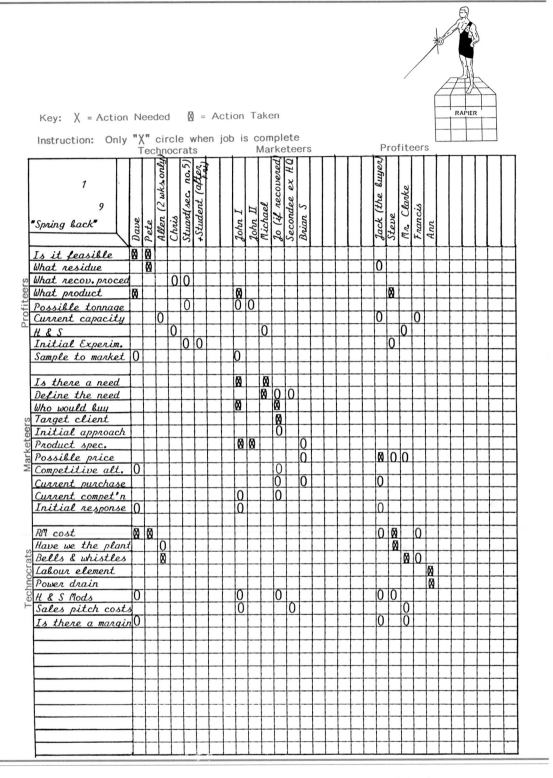

Figure 22.2 Typical RAPIER model deck

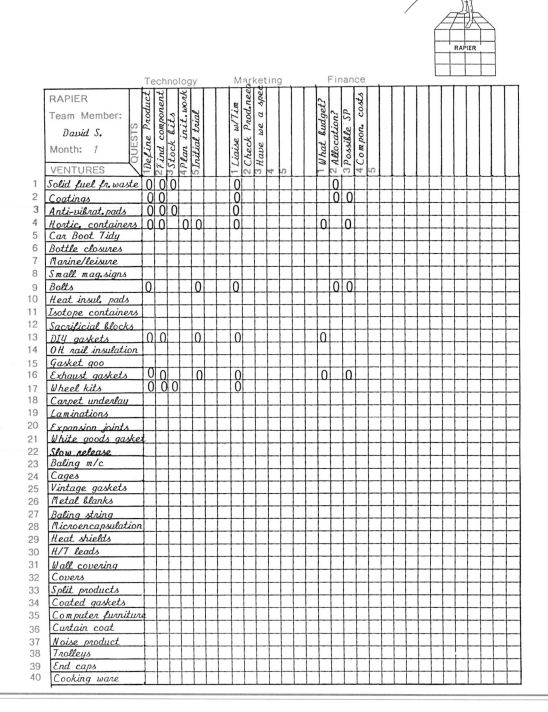

	RAPIER Team Member: David S. Month: 1	QUESTS	Technology					Marketing					Finance																					
			1 Define Product	2 Find component	3 Stock kits	4 Plan init. work	5 Initial trial	1 Liaise w/Tim	2 Check Prod.need	3 Have we a spec	4	5	1 What budget?	2 Allocation?	3 Possible SP	4 Compon. costs	5																		
	VENTURES																																		
1	Solid fuel fr. waste		0	0	0			0					0																						
2	Coatings		0	0				0					0	0																					
3	Anti-vibrat. pads		0	0	0			0																											
4	Hortic. containers		0	0		0	0	0					0		0																				
5	Car Boot Tidy																																		
6	Bottle closures																																		
7	Marine/leisure																																		
8	Small mag.signs																																		
9	Bolts		0			0		0					0	0																					
10	Heat insul. pads																																		
11	Isotope containers																																		
12	Sacrificial blocks																																		
13	DIY gaskets		0	0		0		0					0																						
14	OH rail insulation																																		
15	Gasket goo																																		
16	Exhaust gaskets		0	0		0		0					0		0																				
17	Wheel kits		0	0	0			0																												
18	Carpet underlay																																		
19	Laminations																																		
20	Expansion joints																																		
21	White goods gasket																																		
22	Slow release																																		
23	Baling m/c																																		
24	Cages																																		
25	Vintage gaskets																																		
26	Metal blanks																																		
27	Baling string																																		
28	Microencapsulation																																		
29	Heat shields																																		
30	H/T leads																																		
31	Wall covering																																		
32	Covers																																		
33	Split products																																		
34	Coated gaskets																																		
35	Computer furniture																																		
36	Curtain coat																																		
37	Noise product																																		
38	Trolleys																																		
39	End caps																																		
40	Cooking ware																																		

Figure 22.3 RAPIER personal timetable sheet

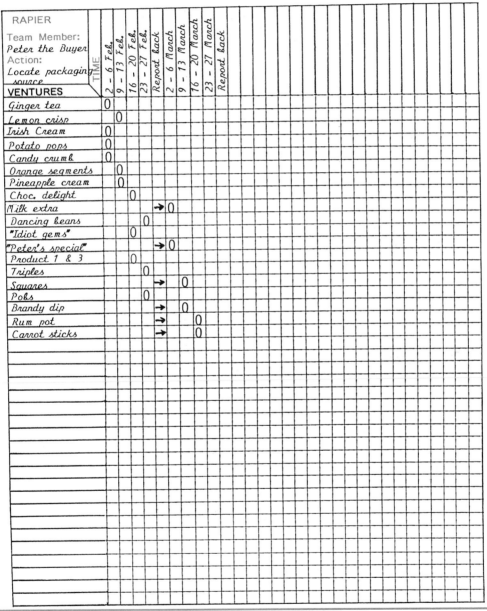

Figure 22.4 Specific action chequer sheet

on the chequer sheet. We have found that this is a useful double-check and can quite frequently pinpoint necessary actions which would otherwise have been missed.

Team member chequer sheets

By taking slices through the model in the vertical plane we can identify specific chequer sheets for specific individuals. Along the two axes of these two chequer sheets are set out all the ventures and all the actions. We produce one such personal chequer sheet for each team member. On these sheets the team member logs the activities taken against all the different ventures in terms of all the different and necessary actions. This then becomes a personal record of individuals' involvement in the ventures and also constitutes a 'time sheet', such as shown in Figure 22.3.

These time sheets often have the effect of cutting out unnecessary debate and speeding up the progress on ventures.

Specific action chequer sheets

By taking lateral cuts through the model we can define a third set of chequer sheets. Each of these relates to a specific action. Along the two edges of these two chequer sheets are set all the ventures and all the

time resource available to RAPIER. Figure 22.4 shows an action chequer sheet.

The chequer sheets also save time. By agglomerating individual actions together it is usually possible for a team member to carry out the same action on a whole range of ventures at the same time, thereby saving time. A typical example is raw material sourcing for a range of ventures. The 'buyer' can tackle all these together. They are also useful in comparing and contrasting the outcome of a specific action *vis-à-vis* a range of ventures. This provides the basis for the assessment and selection procedures which will take place on RAPIER ventures as they develop.

Concluding comments

RAPIER models perform a different purpose from other models used at other stages in our innovation process. RAPIER models are built and used for programming and progressing ventures during the development stage. As such they provide a systematic framework for our multi-disciplinary concurrent approaches.

We have found that this use of modelling can save literally hundreds of team days by avoiding unnecessary duplication and by providing for simultaneous action planning on a range of individual ventures.

23

Modified psychology for venture development

'We've got to get our thinking right about venture development!' In this chapter we present the creativity systems that we use during RAPIER programmes.

The need for a different psychology

At any and indeed every stage of venture development there will be difficulties to overcome. These difficulties are compounded by that part of human nature which resists change. Every step in venture development involves doing something different, something new. These different activities are irksome to anybody whose aim in a company is to maintain the status quo. Most practising managers spend a high proportion of their lives 'fire fighting'. Their natural reaction to something new is to try to avoid it at all costs because it will be a further disruption. To say the least, this will slow down progress on new ventures unless the psychology of the situation can be changed.

Because, in RAPIER, we are striving for a cost-efficient approach, it is inevitable that team members will have to delegate actions to subordinates and other individuals in the organizational structure. Those to whom tasks are delegated will also be very busy individuals and their natural reaction will be anything but positive towards the extra workload associated with new ventures. Again, therefore, there is a real need to address ourselves to the psychology of the situation.

It is imperative at the outset to obtain a definitive statement from the company's board to the effect that tasks associated with RAPIER are regarded as vital to the company's future.

Supportive psychology

What is needed is a psychology within which a supportive approach predominates. To achieve this the rank and file of a company must appreciate the need for the endeavour. This involves giving all those individuals who may be involved in RAPIER either directly or on a delegative basis, understanding of the need for change and the need for business renewal through new ventures. Only through understanding are they likely to be supportive. Individuals who do not understand the need for the RAPIER ventures and who are busy maintaining the status quo should not be part of the team.

We have used the left brain/right brain hypothesis extensively and successfully in this role. It provides a basic appreciation for the need for non-aggressive non-argumentative constructive support.

Creative time saving

Again it must be said that even with a supportive team and a team of supportive delegates to carry out the tasks involved in

139

RAPIER, there will be difficulties. The principal problem is time. However, we have found that creative approaches can be useful in both speeding up the RAPIER process and carving out niches of time within which to carry through the activities. By taking creative approaches to the RAPIER actions, work can be redefined to occupy less time.

This creative redefinition of the tasks is of paramount importance to RAPIER, for this takes ventures through to launch in between a fifth and a third of the conventional time. What is being suggested is that, using a concurrent multi-disciplinary approach, the task can be redefined in a far more time-efficient manner. Companies provide numerous examples where time honoured practices are terribly time wasting! Frequently we have found situations in which a specific task had always been the responsibility of a specific department. Yet on analysis it was found that this department was at arm's length from the necessary interface to carry out the task. We have also found numerous situations where the old adage, give the job to the expert, had the effect of slowing down the system. Conversely, we have had many successes where work sharing enabled the job to be completed far faster than would otherwise have been the case.

All of this, though, requires a much more open-minded approach than usually pervades rigidly departmentalized company structures. Here the multi-disciplinary nature of the RAPIER team is the guiding principle.

The easy way and the harder way

Venture development can be a laborious process but it is usually possible to take some creative short cut.

In business and industry managers generally spend much of their time in meetings and discussions which, on analysis, cannot be regarded as time efficient. They tend to devolve into cyclical argument, point scoring and devil's advocacy. We believe that many such meetings can be cut short by a more rigorous and creative approach to agendas. It seems to us that most business discussions attempt to tackle more than is realistic. Nearly always there are many interrelated aspects within the discussion.

Yet, by standing back from the discussion it is usually possible to spot the key feature, attribute or question which should be the focus of the discussion. Once this has been identified all other matters become subordinate – so less people need to be involved in the meeting.

Key issue creativity

The technique we have evolved for cutting through what would otherwise be rambling contentious discussions, is a structured approach to a form of brain writing. In this we regard the problem as a closed door. The aim is obviously to get through the door to the other side. To achieve this a sequence of steps is necessary. The approach is set out in Figure 23.1.

Key issue sheets are used many times during the RAPIER programme. Something of a learning curve is involved in using this approach but once team members have used it several times it seems to become part of the corporate ethos. Thereafter, whenever a problem arises during RAPIER (and for that matter in many other management endeavours) key issue sheets will be pulled out of drawers and used.

Once familiar, key issue sheets can be used quickly. In this way problems which might have taken a three-hour debate in an unstructured meeting can be resolved in five minutes. However, they can be used only provided the team of people using them is *au fait* with the aims, objectives and psychology of RAPIER.

KEY ISSUE SHEET

Discipline: *Technology*

Venture No: *34* : *Synthetic gemstone "Objets d'Art"*

1. The Quest was: *for a Blue John substitute - a blue/purple lustrous translucent gemstone*

2. The Problem is: *the colour of such materials as Blue John and Amethyst have taken aeons of time to form (geologically)*

3. Re interpretations:

4. Unlock possible solutions

5. Improve the Key

Speed up the geological process

Refire mix in alternative atmosphere

Use a less complex matrix material

Use a simple tinted glass

Make a synthetic matrix

Use Cobalt to colour

Use a simple glass as a vehicle

Use a melt to speed up the process

Use a known colourant e.g. Cobalt Blue

Flux over base material with glass cullet

Add Cobalt Oxide to the melt

6. The best approach is: *Add blue glass cullet to our basic material and refire*

7. Agreed Output: *New series of experiments to incorporate a pre-formed blue pigmentation and glassy matrix by co-firing our base with coloured soda glass cullet.*

Figure 23.1 RAPIER key issue sheet

JIG SAWS : Venture No: 26 : _Microencapsulated Cleaner_

Interlocking of Preliminary Investigations

Marketeers	Technocrats	Profiteers
1. What we need to offer is a solvent that _doesn't_ evaporate.	1. The proposed blend has a vapour pressure less than water and a boiling point of 215 °C	1. The main solvent is secure – we control a permanent source
2. It must be acceptable on Health & Safety grounds	2. We have searched the literature and found three similar compositions cleaned for _food_ usage	2. If we had to change we have two similar sources available – both "food grade"
3. We can't simply copy the PQR product	3. Our blend is a different "composition of matter" and we believe it could be patentable	3. Our sources don't and can't supply PQR – nor could PQR claim we have "pinched" one of their sources
4. It must give us a margin at the market ruling price of £10/litre	4. Our RM costs are still less than £1/litre and blending is on existing plant – it should give a good margin	4. The Preliminary Costing suggests we should clear £6/litre gross margin. Even if we have to source from ZX Ltd. we would still clear £5/litre
5. It won't sell if it is at all noxious to use – the "Cleaners" are powerful – viz. the Shop Union	5. We've sampled it to your particularly fractious client in Sheffield and their work force actually said they liked it! (No! before you ask, it's not addictive)	5. At the price you are proposing to offer it – it should be addictive. ?Could we go for £12/litre and drop back if competition arrives?

Figure 23.2   RAPIER jigsaw sheet

Multi-disciplinary matching

Key issues are best addressed by small teams of individuals from within the RAPIER team. We have found that a team of three seems to be the optimum size. Above three the debate extends, indeed we wonder whether there is a fundamental law that 'the rate of progress in a debate is inversely proportional to the number (of people) debating'.

Because of the nature of key issues the groups used to debate them could tend to be uni-disciplinary. Here lies a danger that the solution found by an individual discipline will not match the perception of the venture by other disciplines. Typically this can give rise to a mismatch between the product as perceived by the technocrats and as perceived by the marketeers. If this is not remedied the result will be sub-optimal.

Jigsaws

The way we remedy this potential mismatch in RAPIER is to use 'jigsaw sheets'. These are very simple three-column sheets in which the venture is redefined after each key issue is resolved. The aim is to make sure that the definitions of the venture by the three individual disciplines within the RAPIER team are matched. A fairly typical example of a jigsaw sheet is shown in Figure 23.2. It is sensible to use a pencil when filling in jigsaw sheets so that minor modifications can be made to the definitions in the columns, thereby refining the match.

Concluding comments

By using simple derivatives of the brain writing technique such as key issue sheets and jigsaws, we have found that a far more open-minded and less antagonistic ethos can be built during venture development. This stands ventures in good stead in terms of arriving at launch with a matched perception of need, means and opportunity.

By avoiding internal strife, for example rambling cyclical arguments between the disciplines, a great deal of time can be saved during venture development. Progress can be monitored using RAPIER models and it is quite surprising how quickly ventures are developed to a point where a decision can be made to launch them.

24

A format for venture development

In this chapter we present our interlocking format for a systematic and yet creative approach to the difficult task of venture development.

Introduction

We have two interlocking formats of the RAPIER programme for venture development, a six-month and a one-year version, incorporating both modelling and the modified psychology that we have set out in the two preceding chapters. Again, either the model or the modified psychology can be used as a free-standing management option. However, we have found that by interlinking these two techniques together with other supportive techniques a very rapid, cost-effective approach to venture development results.

An outline of RAPIER

The RAPIER programme for rapid, cost-efficient venture development is a consultancy package of the process type. As such it is a collaborative programme using a concurrent multi-disciplinary approach to new business development. The role of the consultant in RAPIER is process provision, process control and co-ordination, process monitoring and, where necessary, process

arbitration – very much the role of the outsider or the amiable amateur. An external chairman of the team can frequently ask those questions which go beyond the mind sets of the corporate structure.

The RAPIER team is, above all else, a balanced multi-disciplinary team. Typically we use twelve-person teams:

- four marketeers
- four technocrats
- four profiteers

The team must be committed to venture development and the time they spend on venture development must be authorized by their departmental heads. The normal time allocation for RAPIER is one day a week per team member throughout the six months or the year, as the case may be.

The task of the RAPIER team is to take a portfolio of ventures which have emerged from SCIMITAR or BLADE through to capital expenditure proposal and onwards to launch. Depending on the type of industry the six-month or one-year programme will be targeted to produce either capital expenditure proposals or actual launch. Depending on the normal gestation period within the industry, it may require a year to reach capital expenditure proposals. However, we have found that RAPIER can certainly halve and on occasions quarter the conventional development time taken in sequential approaches.

144

THE RAPIER PROGRAMME

Month	Day	Session	Item
0	0	1	Listing potential Ventures (ex SCIMITAR etc.)
		2	Listing all Quests that must be answered
		3	Resource Allocation
		4	Action Plans - Preliminary Investigation
1	1	1	Non-assessive Concept Development
		2	Preliminary Investigation Results
		3	Initial Selection
		4	Action Plans - Feasibility Study
2	2	1	Non-assessive Concept Development
		2	Feasibility Study Results
		3	Second Stage Selection
		4	Action Plans - Field Trials
3	3	1	Non-assessive Concept Development
		2	Field Trials - Results
		3	Third Stage Selection
		4	Action Plans - Product Optimisation
4	4	1	Non-assessive Concept Development
		2	Product Optimisation - Results
		3	Fourth Stage Selection
		4	Action Plans - Viability Study
5	5	1	Non-assessive Concept Development
		2	Viability assessment - Results
		3	Fifth Stage Selection
		4	Action Plans - Actionability Study
6	6	1	Non-assessive Concept Development
		2	Actionability Study - Results
		3	Venture Proposal - in Written Form
		4	Presentation and Overview

Figure 24.1   The RAPIER programme

The RAPIER team is trained to reuse the programme without any further input from the consultant. In this way a continuum can be established which in turn will lead to consistent and continuous new business growth.

RAPIER teams reduce the conflict within new business development activities. Although some conflict is inevitable, and in healthy companies its resolution leads to improvement, we argue that conflict should be minimal. There should be little friction between departments because the team is authorized, resourced, seconded and reports back to the established departmental structure of a company. However, we have found that the RAPIER approach has the effect of blurring departmental interfaces to the point where other corporate activities can be carried forward on a simultaneous multi-disciplinary basis.

The programme

The RAPIER programme revolves around a series of monthly plenary meetings in which all team members are present together with the consultant's team. A typical programme is outlined in Figure 24.1.

Between these plenary sessions action plans are carried out in a multi-disciplinary three-dimensional array, established, controlled and monitored using a three-dimensional model. The significant psychological changes involved in management approaches within RAPIER are handled using key issue sheets and jigsaws. These require a fair degree of creativity and open-mindedness. Here again the role of the amiable amateur or outsider is most important: companies which have used both the modelling and creative approaches to RAPIER without an external influencer have struggled!

Output

Towards the end of the RAPIER programme ventures are sufficiently developed to be presented to the company's board for a decision to launch. Typical output from RAPIER is five such ventures, which are written up as Capital Expenditure proposals. We have evolved a standard format which is acceptable to the vast majority of companies. This standard is set out in Figure 24.2.

INDEX

1. EXECUTIVE SUMMARY

2. ORIGIN OF CONCEPT

3. TARGETED DATA OF RAPIER EXERCISE

4. PRELIMINARY INVESTIGATION

5. FEASIBILITY STUDY

6. INTERLOCKING OF MULTI-DISCIPLINARY ASPECT

7. PRACTICAL VENTURE DEVELOPMENT

8. ACTIONABILITY STUDY

9. BUSINESS DEVELOPMENT ACTIVITIES

10. NEW BUSINESS UNIT DEFINITION

11. NEW BUSINESS OPTIMISATION

12. LAUNCH DEFINITION

13. PREPARATION FOR TRIAL LAUNCH

14. PROPOSAL

15. CONCLUDING COMMENTS

Figure 24.2 RAPIER capital expenditure proposal format

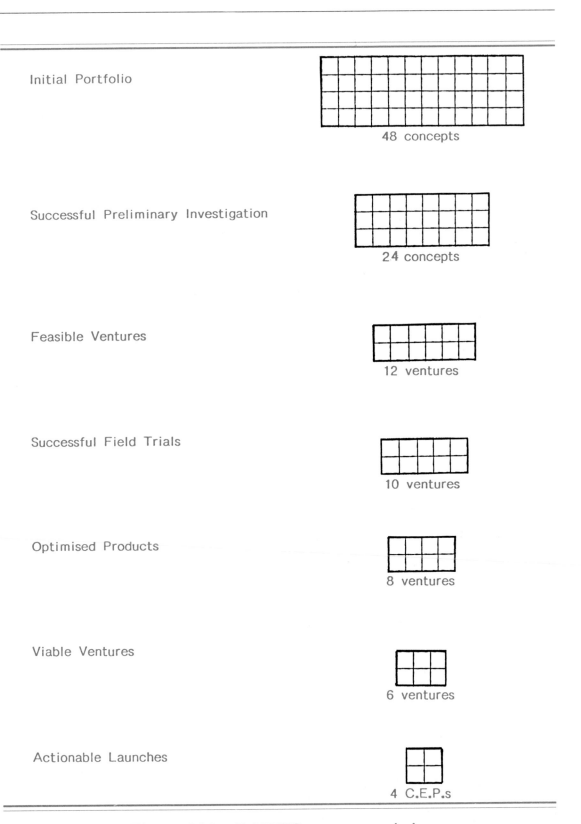

Initial Portfolio

48 concepts

Successful Preliminary Investigation

24 concepts

Feasible Ventures

12 ventures

Successful Field Trials

10 ventures

Optimised Products

8 ventures

Viable Ventures

6 ventures

Actionable Launches

4 C.E.P.s

Figure 24.3   RAPIER output statistics

Capital expenditure proposals are presented both orally and in written form to boards at the end of the RAPIER programme.

Statistics and case studies

Although we have been running RAPIER programmes since the early 1980s they are still among the most difficult we offer and we certainly acknowledge that venture development is one of the most problematical stages in the innovation process.

Typical statistics for a RAPIER programme are set out in Figure 24.3. It must be stressed that it is most important to go into RAPIER with an adequate number of possible ventures. We prefer a figure nearer fifty than twenty to allow for fairly rigorous assessment at the four selection stages during the programme.

In over 75 per cent of RAPIER programmes we actually see first orders taken during the programme. On average we present five capital expenditure proposals to company boards at the end of the exercise, one of which will have reached a first order situation.

In terms of the cost-efficient usage of time we regularly achieve five capital expenditure proposals, including one which is already selling, from a total time commitment of 300 team days. We believe that this is at least three times faster than using conventional approaches to venture development.

A major agrochemical and food company was at first incredulous about the following idea derived from their SCIMITAR Programme:

The concept of using an extract of the territory marker of a predator to ward off animal pests in agriculture.

The question was "how do we handle such a concept?" The answer was to keep their SCIMITAR Team together and run this venture (together with approx. 20 others) through to a full scale C.E.P. presentation, on a concurrent multi-disciplinary basis, i.e. RAPIER.

The Team met on a circuit of company sites on a monthly basis. Between the meetings, each Team Member tried to allocate one day per week (it is admitted that the actual time committed was less).

Progress seemed gradual to the Team but was rapid when viewed from the outside.

Six months later, this concept had survived RAPIER preliminary investigation, had had successful field trials, in which it was shown that deer can be prevented from de-barking sapling firs if the saplings are sprayed with extracted marker of a lion.

By the twelve month point, the company was poised to launch a new product into the forestry sector, based on a synthesised analogue of the territory marker of a lion.

Figure 24.4   RAPIER case history

By way of a case example a RAPIER programme conducted in the early 1980s is set out in Figure 24.4. We have chosen this case study to indicate how truly novel concepts can be progressed rapidly using RAPIER.

Concluding comments

By using an interlocking programme including both a model-based systematic approach and a modified psychology, many of the pitfalls associated with venture development can be obviated. The guiding principle of our RAPIER approach is that, because business is multi-dimensional, venture development teams should be multi-disciplinary. This allows for concurrent rather than sequential development of ventures, which in turn yields massive savings in resources consumed and time to launch.

PART 7

NEW BUSINESS LAUNCH AND COMMERCIALIZATION

CREATIVE

UNIFYING

TECHNIQUES FOR

LAUNCH

APPROACHES,

START UPS AND

SALES

CUTLASS

There are many case studies of ventures taken through to launch which ultimately failed in the commercialization stage. These make sad reading when the wastage of resources is considered. In this final operational part of the manual we present new management approaches to these vital early commercialization situations.

25

An introduction to launch and commercialization

In this chapter we examine the difficulties experienced in launching industrial and commercial ventures and suggest the need for alterations to the conventional approach.

Starting point

At the end of the RAPIER exercise the necessary venture development will have been carried out so as to prepare new businesses for a decision to commercialize. Launchable ventures will have reached this stage by two different routes:

1 Organic growth concepts which have been produced as ideas and developed in house.
2 Externally sourced concepts which have led to licences or possible acquisition opportunities.

In either case the starting point for the CUTLASS exercise is the company board passing a capital expenditure proposal.

'Go for it'

Because time is of the essence it is obviously sensible, as soon as the board has made its decision, to launch the venture as quickly as possible. Only in this way can the lead time generated by the development

process be converted into a 'honeymoon period' in the market. If delays occur at this stage the competition will inevitably catch up and the 'novel product' will suddenly become a 'me too'.

Approach

Launch and commercialization are the important but complex finales to the process of innovation. Many pitfalls can destroy ventures, even at this late stage, and in terms of commitment and spend the launch and commercialization stages are the most costly. We recommend an interlocking multi-disciplinary and unifying approach to commercialization. Our main concern is that ventures will arrive on launch day with several vital components missing. It is all too easy for the marketeers' concept of what they are about to launch to be subtly different from that which the technocrats are about to make!

For this reason the vital word in the CUTLASS acronym is 'Unifying'. It is paramount to unify the various dimensions of business at the launch and commercialization stage.

The CUTLASS team

Because of the need to unify the dimensions

of the new business unit to be launched we believe that a multi-disciplinary team should be in charge of launch and commercialization. The CUTLASS team can therefore be an extension of the RAPIER team. Most of our clients do in fact take the RAPIER team into the launch situation. However, the RAPIER team should be enlarged and strengthened in certain key areas (by a process of secondment) for the period of launch.

The task

The task of the CUTLASS team is to take ventures on which capital expenditure has been agreed through to launch and commercialization. This task is truly multi-disciplinary: it involves making sure that the product is ready for launch, the market is ready to accept it and that the yield from the venture is financially acceptable.

When confronted with a multi-dimensional task our standard response is to adopt a modelling approach to focus the process. In the next chapter we set out the modelling system used for CUTLASS, and in Chapter 27 we present the creative approaches that we use to underpin the model. Finally, in Chapter 28 we provide a composite format for CUTLASS.

26

Systematic approaches to launch and commercialization

At the point of launch all six dimensions of business are equally important. Six dimensions can be very difficult to envisage; CUTLASS models act as a simplifying device and help to 'unify' all aspects of a venture.

Dimensions

Because we contend that in CUTLASS all six dimensions of business are important, it has been necessary to devise a six-dimensional model. It may be thought difficult to build a physical model in more than three dimensions, but we have evolved a system, shown in Figure 26.1.

CUTLASS Cubes

This multi-dimensional approach is achieved by using cubes – square-sided cardboard boxes. The six faces are capable of carrying six-dimensional information. The information we ascribe to the faces of the box is:

- resources consumed
- processes applied
- markets serviced
- finance involvement
- people involved
- timing

Model building

CUTLASS cubes allow us to build dynamic and informative models. We use relatively large boxes (8″ × 8″) sealed shut. These carry six sheets, on each of which is written all the currently available data related to a specific dimension of the venture.

CUTLASS cubes are easy to handle, they can be passed round the team and added to and they are a multi-dimensional vehicle for creative debate.

Output of CUTLASS models

The cube faces provide a highly systematic approach to comparing available data against the norm or targeted data needed at each stage of the commercialization process. Figure 26.2 gives an example. Information is updated regularly, simply by adding new sheets to the cube's faces.

By comparing and contrasting the information present, gaps in thinking and understanding become apparent and can be plugged before they critically affect launch and commercialization.

The output from the modelling exercise, which takes place weekly during each plenary session of the CUTLASS team, is a

A simple cubical (i.e. square faced) cardboard box

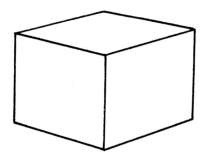

has six faces, and each can be regarded as a dimension. Venture data can then be collated for each of the six dimensions of business and mounted on the six faces:

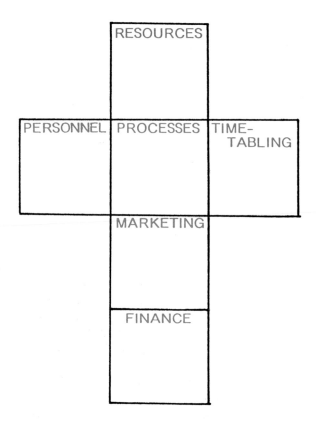

The CUTLASS Cube looks like this if opened out. (Normally the box is taped shut).

Figure 26.1   CUTLASS cubes

VENTURE: *Shielding Paint*	DIMENSION: *Resources/RMs*	UPDATE: *Week 23*

Item: *Version for manual spraying*   *Liaison*

1. *Need to reformulate to remove chlorinated solvent for . . JG/VA manual sprayer*

2. *Source of non-chlorinated solvent found . . . . VA*

3. *Price of non-chlorinated solvent . . . . VA/WK*

4. *Trial to check for settling etc. . . . . TS/VA*

5. *H & S leaflet modified . . . . .VA/WK/CD*

6. *Label modified (note new colour code) . . . . .VA/CD/Printers*

7. *Check effect on shelf life . . . . TS/JG*

8. *Re-work costing . . . . WK/JG*

9. *Launch (date target) . . . . Week 29 . . . . . . JG/Salesmen*

10. *Buy new solvent stocks . . . . . . Week 26 . . . . VA/TK*

11. *Trial run . . . . Week 27 . . . . TK/VA*

12. *In house verification . . . . . . Week 28 . . . . TK/JG*

Figure 26.2   CUTLASS cube face

sequence of action plans addressing gaps spotted on the faces of the venture cube. These action plans are multi-disciplinary and the interactions between disciplines can be checked out.

Concluding comments

CUTLASS models are theoretically the most complex of all the models that we use throughout the innovation process yet they are simple to use in practice. Because it is very important to unify all the dimensions of business at the point of launch it follows that the model for CUTLASS will be six-dimensional. By using CUTLASS cubes it is possible to build, into three real dimensions, dynamic information on all the six dimensions of business.

27

Creative techniques for launch and commercialization

In this chapter we suggest that the normal pscyhological approach to launch and commercialization of business and industrial ventures is sadly lacking. And we provide some alternative creativity techniques which can improve the cost efficiency of launch and commercialization.

Mental attitudes

'A company is a group of loyal warriors battling for a common victory.' When we suggest that this should be the definition of a company it is often greeted with laughter. Most businesses and industrial concerns seem to engender far less loyalty. Throughout the innovation process we see battles breaking out within the organizational structure. It can be argued that for as long as the company's ventures remain internal the battle damage is limited. However, at the point of launch ventures become public and it is highly damaging if these battles are seen by outsiders. We therefore argue that at launch everybody involved in a venture must be loyal to it.

Conventional difficulties

In the launch and commercialization process there are many common pitfalls and problems. Many of these stem from the fact that departments within companies do not adhere to corporate strategy. There are normally inter-departmental barriers which allow for a dichotomy of aims and objectives. It is very rare to find all company departments truly loyal to the venture as the launch date approaches: it is far more usual to find situations in which one department blames another for shortfall in performance.

We have even found ventures launched as a 'points scoring exercise' by managements who know full well that the input from certain key departments is incomplete and that the venture was being put at risk.

Departmental wars – and how to stop them

We use a creativity technique based on CUTLASS modelling to focus and unify the activities of all departments involved in a launch.

If the updated information on the six faces of the cube is in tune it may fairly be assumed that all six dimensions of business are equally developed, interlinked and unified. If this is not the case it is possible for work to be commissioned on specific items or dimensions while at the same time noting implications of change on other dimensions. This is not as complex as it may seem. Although several of the six dimensions may interact it is usually possible to see the dimension or dimensions which are subject to change and therefore require further development.

Creative communications

The key to preventing disagreements and departmental warring is to increase the effectiveness of communications – and the CUTLASS cube seems to achieve this. However, the basic ground rules for creative non-assessive debate are most important at this stage. It is not useful to act as a devil's advocate or as an inquisitor in these situations. What is required is a general recognition of the corporate mission and a specific recognition of the need to launch successfully the venture on which a capital expenditure proposal has been agreed. Under these circumstances, irrespective of previous differences between departments, it is very important that all groups pull in the same direction. Difficulties will inevitably arise but if these can be discussed creatively, that is, non-aggressively, solutions can be found and the team will be strengthened. By passing the CUTLASS cube from hand to hand around the team, creative unifying communications are established. So, at the CUTLASS stage our systematic and creative approaches become one in the cube.

Again, as so many times throughout the innovation process, we see a role for the amiable amateur from outside in controlling the debate, creating specific focuses and helping in discussions moving towards a workable solution.

Difficulties

The difficulties that may be experienced in launch and commercialization are legion; we list some of the more common in Figure 27.1.

Anything and everything can go wrong as the days tick away towards the launch. By this time of course the launch date itself will be indelibly inscribed on the calendar. Funds will have been spent on advertising, promotion and publicity – in all probability some kind of 'event' will have been scheduled for that date. It may be as simple as a first introduction to a major client or as grand as a TV campaign. Unless it is absolutely impossible to adhere to that date the aim of the team must be to launch on that day.

Although there are many difficulties, if a systematic approach to the innovation process has been taken discipline will prevail, and if a creative psychology has been developed, an *esprit de corps* should have been built around the team. If this intangible is in place there is a reasonable chance that everybody will indeed pull in the same direction and difficulties will be ironed out.

Ink drying on an order

The old saying that 'success breeds success' is never more true than in launch and commercialization. Once the initial difficulties have been overcome then, although the team may have been exhausted by its efforts, there will be a moment of sheer elation when the first order is taken. All the difficulties which have been faced by the team then fade into insignificance and the team can transcend any future difficulties – at least in terms of psychology. One cannot over-exaggerate the importance of an initial order. (An initial order, achieved prior to the CUTLASS scenario, is a wonderful boost for the venture.) Once the ink starts to dry on an order the team's elation can easily be translated into even more positive action. Just as the winner of a race will happily jog around the circuit yet another time, so an innovation team will happily spend further hours at the end of a long day checking

---

1. A common error is to go for launch before the product is ready. This brings forward a host of criticisms and rejections from the market place.

2. Launching when marketing department is insufficiently prepared is another common error. This again gives rise to major problems in the market place.

3. Failing to resource the build up of sufficient stocks prior to launch is also a fairly common phenomenon. This has the effect of titillating the market and then cutting off the flow.

4. Plant breakdowns and difficulties are a common phenomenon around launch time. New products, whether they are going through new plant or existing lines, frequently stretch and test engineering efficiency.

5. People problems often arise at launch and commercialisation. Typically we find that people's time has not been recognised as a precious resource and insufficient time has been dedicated to launch or commercialisation.

6. Another surprisingly common problem is to find that a vital raw material for the new venture dries up very shortly after launch. This can of course be due to devious activities by competitors.

7. Pricing policies often come unstuck immediately after launch as competitors drastically undercut ruling prices in an attempt to squeeze you out.

8. Price is not the only retaliatory weapon that competitors can use. There are a whole host of semi-legitimate perks which can be thrown into the market so as to confuse potential buyers at or around your launch date.

9. Quality is of supreme importance immediately after launch. And, at this stage, it is quite usual to find that because of minor teething troubles in the plant etc., quality suffers.

10. Availability and consistency of supplies including an efficient distribution system are also of vital importance during launch and commercialisation and frequently come unhinged.

Figure 27.1   Specific CUTLASS difficulties

every circuit in the kit they are about to sell. This of course maximizes the chances of the order becoming an invoice – and then a cheque!

The show is on the road

Once a semi-continuing stream of orders starts to appear the show is really on the road. Of course there will still be difficulties: a mismatch between production capability and marketing requirements, perhaps, or friction between production, marketing and finance. It could well be necessary on occasions to cut margins to establish a position in new market sectors deemed to be important. Again at this stage the team spirit which has been built over the months, or indeed years, in terms of a multi-disciplinary approach to innovation will be of benefit.

There will be times, there always have been times, when one discipline comes under specific and intense pressure. In many commercial and industrial situations this is an opportunity for conflict. A department or function which is seen to be labouring under the pressures of launch and commercialization is often criticized within conventional approaches. In the CUTLASS approach the recognition that a particular department is suffering unduly, is a trigger for all other departments which are involved to spring to their aid. Again, we see a blurring of departmental interfaces. This may well sound far too egalitarian for the average business structure, yet it happens every day in every industry. It happens more in smaller entrepreneurial organizational structures, and that is why they are often better innovators. However, CUTLASS can create a unified band of 'warriors' within a large warring throng.

28

A format for launch and commercialization

In this chapter we present an interlocking programme for the launch and commercialization of new products or services in business or industry.

Introduction

At this, the final step in our innovation process, systems and creativity unify. The CUTLASS cube is both systematic and a creative approach to launch and commercialization. As such it can be used as a free-standing single management methodology. However, we have found that its impact is even greater if it is used as the fulcrum for an interlocking programme.

Approach

Faced with a multiplicity of potential difficulties we use our understanding of business as a six-dimensional complexity, and first build a unifying model. By building a CUTLASS cube and appending all available information on to the six faces of the cube – in six dimensions – it is possible to check out these potentially problematical situations.

Our approach is simply to stick update sheets on each of the faces of the cube. Each face represents one of the six

dimensions of business and each update sheet represents the current state of affairs *vis-à-vis* that dimension in the venture. We present a typical update sheet in Figure 26.2. As will be guessed, it is sensible to use pencil and eraser on these sheets rather than indelible ink!

Programme

Although the physical action of building a CUTLASS cube is in itself both a systematizing and a creative approach, we find that its usefulness is maximized if a programme is built around it. The CUTLASS programme is a sequence of weekly meetings attended by all members of the team involved in launching and commercializing the venture. At the meetings each of the six dimensions is discussed and an update sheet produced. The update sheets are then stuck to the CUTLASS cube, which is then scrutinized in detail.

The juxtaposition of the update sheets on the faces of the cube seems to make it so powerful. Very few people can actually think in terms of six dimensions, yet we have found that CUTLASS teams can see combinations, permutations and potential difficulties arising out of the interaction between the six dimensions of business as displayed on the faces of the CUTLASS cube.

```
┌─────────────────────────────────────────────────────────────────┐
│ VENTURE: Shielding Paint  DIMENSION: Resources/RMs  UPDATE: Week 23 │
├─────────────────────────────────────────────────────────────────┤
│ Item:   Version for manual spraying                      Liaison  │
│                                                                   │
│  1.  Need to reformulate to remove chlorinated solvent for .. JG/VA │
│      manual sprayer                                               │
│                                                                   │
│  2.  Source of non-chlorinated solvent found . . . . . . . . . VA  │
│                                                                   │
│  3.  Price of non-chlorinated solvent . . . . . . . . . . . VA/WK  │
│                                                                   │
│  4.  Trial to check for settling etc. . . . . . . . . . . TS/VA   │
│                                                                   │
│  5.  H & S leaflet modified . . . . . . . . . . . . . . VA/WK/CD  │
│                                                                   │
│  6.  Label modified (note new colour code) . . . . . VA/CD/Printers │
│                                                                   │
│  7.  Check effect on shelf life . . . . . . . . . . . . . TS/JG   │
│                                                                   │
│  8.  Re-work costing . . . . . . . . . . . . . . . . . . WK/JG    │
│                                                                   │
│  9.  Launch (date target) . . . . . . . Week 29 . . . . . JG/Salesmen │
│                                                                   │
│ 10.  Buy new solvent stocks . . . . . Week 26 . . . . . . VA/TK   │
│                                                                   │
│ 11.  Trial run . . . . . . . . . . . . Week 27 . . . . . . TK/VA  │
│                                                                   │
│ 12.  In house verification . . . . . . Week 28 . . . . . . TK/JG  │
└─────────────────────────────────────────────────────────────────┘
```

Figure 26.2   CUTLASS cube faces

It can be argued, and we would accept, that it is mainly the discipline of getting all key individuals into the same room at the same time once a week that makes the CUTLASS programme so efficient. Situations within companies can become very frantic at launch and the early stages of commercialization, and discipline is of paramount importance. Things inevitably go wrong and it is pointless for individual departments to blame others. What is required is a loyalty to the venture at launch and this is engendered by the CUTLASS meetings.

Duration

CUTLASS teams – sometimes called CUTLASS clubs - meet for half a day each week for as long as is necessary. The duration of launch and the early stages of commercialization take varying periods depending on industry, market, product and company. And so it is usually something of an open-ended programme.

Indeed we know of client companies who run on CUTLASS clubs in association with ventures long after launch. The unifying effect of the CUTLASS cube and the team's understanding of the six dimensions of business stand the venture in good stead as it progresses up the S curve and ultimately plateaus out as a mature product.

Finally, we see a role for CUTLASS clubs in ongoing business: to improve communications between individual departments thereby reducing interfacial barriers. The lessons learnt in terms of combining forces and unifying one's battle plan in new business venturing seem to us to be equally applicable in the management of ongoing business.

Commitment and time scale

By using the CUTLASS club approach the cost effectiveness of new business commercialization can be greatly improved. On average a twelve-person team can take a venture through to full commercialization in between six months and a year in industries which would normally require two to three years.

Up to a point, if more people resources are committed to CUTLASS and dedicated to a specific venture, the faster launch and commercialization are achieved. Most companies use one CUTLASS club to launch one venture at a time. In some industries where ventures tend to be relatively small we have found it is possible to use one company-wide CUTLASS club to launch several ventures simultaneously and, over a period of months and years, a whole sequence of ventures. However, we

feel on balance that CUTLASS clubs which are exclusively associated with individual ventures work better.

The ultimate advantage of an exclusive CUTLASS team is that it constitutes the permanent management for the venture in embryo form. All that is required is gradually to increase the status of the members of the club so that they become fully fledged business managers and ultimately the directors of the new business unit – if it grows sufficiently large.

Output and case study

The twin output from a CUTLASS programme comprises a new and profitable business activity launched in a cost efficient manner, and an integrated entrepreneurial team who can manage and develop the business. The bonus is a system for avoiding inter-disciplinary disputes within a company – and that can have far-reaching effects on all business, both established and innovative.

By way of a case study to exemplify the size and type of venture which can be tackled we show in Figure 28.1 the example of a CUTLASS team taking a revolutionary new approach to the manufacture and distribution of a basic commodity through to launch and commercialization in six months.

Concluding comments

Many of the difficulties experienced in launch and commercialization can be avoided and many of the costs reduced by a concurrent multi-disciplinary approach. However, because all six dimensions of business must be 'unified' at this stage you

It is arguably more difficult to create and successfully launch a fundamental breakthrough in a mature commodity sector than in a newer, faster moving industry. Mature industries take decades to change and companies rarely make rapid, massive gains even though they employ large teams and major revenue expenditure on R. and D. etc. . . .

A CUTLASS Team of six individuals was drawn together on a multi-disciplinary basis from within the existing management of a medium sized basic material handling and extraction company late in 1979. Earlier that year an alternative method for manufacturing and marketing agricultural fertiliser had been evolved by a SCIMITAR Team and developed by a RAPIER Team both within a subsidiary company of the group.

The CUTLASS Team were sanctioned to attack the market vigorously and create a significant new position for the group in agricultural fertilisers. At the start of the campaign, the group were only known in the much smaller market of horticultural fertilisers and ranked below at least 50 other companies on a production tonnage basis.

The concept developed switched the point of "manufacture" from major installations to small regional blending plants. This minimised the capital spend that would have been required to break into the market; and each plant was franchised, reducing the group's involvement still further.

The six person CUTLASS Team planned and executed the establishment of six regional plants in six months, the first coming on stream three months into the campaign. By the end of the first six months the venture was producing at a tonnage rate of 200,000 t.p.a. and ranked fourth largest fertiliser manufacturer in the UK.

The difficulties encountered in such a vigorous attack on a conservation market using a novel approach were numerous but the CUTLASS Team approach succeeded.

Figure 28.1   CUTLASS case study

need a six-dimensional management system to help in this complex situation.

When used as part of an interlocking programme CUTLASS clubs provide a powerful approach to management development and action planning during and after launch of innovations.

PART 8

AN OVERVIEW OF THE INNOVATION PROCESS

An overview of the innovation process

In this final part of this manual for the 1990s we set out our basic conclusions drawn from the work presented on our studies of innovation in business and industry.

29

The SWORD approach to business and industrial innovation

In this chapter we give a brief overview of the systems, techniques and programmes which we have developed for the various stages in the innovation process.

Analysis of difficulties – SWORD

Most companies recognize that there are difficulties involved in the innovation process and many feel that their approach is not ideal. Under these circumstances we recommend that the first stage is an analysis procedure to pinpoint the difficulties which are specific to the company.

Our approach to this analysis is to define business and innovation as six-dimensional complexities and to study problems against this format. It then becomes possible to isolate specific difficulties on specific dimensions and to recognize adverse interactions between the dimensions.

Innovative strategic planning – SPEAR

Having recognized that there are difficulties associated with innovation within a particular company the next step should be a reappraisal of available options through a new approach to strategic planning.

Strategic plans need not be mere projections of current experience but can be innovative. By using a three-dimensional modelling approach based on products, markets and time, together with the 'tea cup game' for examining future scenarios, we have found that companies can evolve a far more creative strategic plan.

Business realignment – SABRE

Due to the ferocity of competition and the need to maximize ongoing profits, companies are reluctant to sanction substantial additional expenditure on innovation. However, this need not be a bar to an innovation programme.

We have found that in most companies it is possible to realign business resources and in this way free key individuals for innovation activities. By building a three-dimensional model based on business units, management tasks and management options and backing this up with the 'playing card game', which allows unpalatable options to be studied open-mindedly, it is usually possible to free 10 per cent of indirect resources for innovation.

Idea generation – SCIMITAR

Having freed resources for an innovation

169

programme the next step is to generate ideas to feed into the programme.

The SCIMITAR approach is the oldest and most established of our techniques. It works on the basis of the three-dimensional model involving resources, processes and markets as the axes (or derivatives) together with the 'creativity club' technique, which is a hybrid of brain writing, attribute listing, lateral thinking and synectics. This programme can be relied upon to generate literally hundreds of relevant new business ideas for the company.

External sourcing of innovation – BLADE

Not every company would have the resources to develop and launch internally generated new business ideas. An alternative is to capture external innovation opportunities. This involves the 'courtship approach' to licensing and ultimately acquisition.

The three-dimensional model built in BLADE uses as its axes product, market and source, where source is the external company to be courted. This model is then backed up by the 'four leaf clover' creativity technique intended to remove mind sets about which forms of business are acceptable to a company. BLADE will generate hundreds of licensing and acquisition possibilities, thereby offering the luxury of choice.

Venture development – RAPIER

Having generated, assessed and selected innovation options the next stage is to develop these as ventures.

The RAPIER programme uses the three-dimensional model based on resources, processes and ventures, and this is backed by creativity techniques such as 'jigsaws' and 'key issue sheets', which aim to avoid the pitfalls frequently encountered in the difficult process called venturing.

Launch and commercialization – CUTLASS

The final stages of the innovation process are launching the new venture and commercializing it. The CUTLASS approach is a unifying technique to ensure that all facets of a venture are fully understood when it is launched. At this stage all six dimensions of business come together in the 'CUTLASS cube' model and creativity approach.

We have found that the lessons learnt during innovation of new business can be carried forward so that they change the fundamental operational approach of a company, making it far more creative and less antagonistic.

30

Conclusion

In this final chapter we draw together our concluding comments on the innovation process and set out some guiding principles for companies in the 1990s.

Business and innovation

For the foreseeable future developed nations will continue to earn their way in the world through business and industry.

The rate of change in business and industry seems to be accelerating. This is typified by product lives which seem to be shortening in most industries. There is also a gradual trend towards more service-orientated business as against primary manufacturing. Or perhaps it should be said that the geographical focus for primary manufacturing is moving away from advanced nations towards developing nations.

Given this rate of change innovation will be of paramount importance. Innovation is the management of change in terms of business and industry.

The management of change

Through working programmes for innovation on a consultancy basis with a wide range of client companies we have come to believe that the management of change is far more difficult than the maintenance of the status quo in business and industrial terms.

Indeed we have argued in this book that practising managers need to learn to use a powerful new arsenal of weapons for the efficient management of change, that is, innovation. In this book we have presented a range of such weapons, both systematic modelling approaches and creative techniques. Many of the systems are significantly different from those practised in the management of existing business and many imply a changed psychology.

Business and industrial psychology

We believe that in the twenty-first century business and industrial psychology will be of paramount importance. Of all the six dimensions of business we believe that it is the people axis which is of crucial importance. Within virtually every business or industrial organizational structure we find individuals whose psychology resists change. In every other respect such individuals are adequate corporate people. However, standing in the way of progress and innovation they can be extremely damaging. For this reason we believe that a new science will be born in the twenty-first century called 'Innovation psychology'.

Creativity and innovation

Creativity is not about systems, it is an attitude of mind. The great majority of managers in business and industry *can* be creative, but they need to be encouraged and the right psychological framework must be provided.

Innovation is a process. It involves creativity but it must also involve disciplining systems to apply that creativity in a focused way so as to generate relevant new business options. Innovation is therefore a balance of psychology and systems.

In this book we have tried to present a balance between psychology and systems. Only if this balance is right can a business or industrial company rely on innovation for growth during the 1990s and beyond.

BIBLIOGRAPHY

Abetti, P. A. (1983) 'The Process of Technological Innovation in Large and Small Companies', *American Chemical Society Reprints*, August.

Ackoff, R. A. and Vergara, E. (1981) 'Creativity in Problem-Solving and Planning: A Review', *European Journal of Operational Research*, Vol. 7, pp. 1–13.

Adair, J. (1984) *The Skills of Leadership*, Gower, Aldershot.

Beckhard, R. (1969) *Organization Development: Strategies and Models*, Addison-Wesley, Reading, Mass.

Belbin, M. (1981) *Management Teams: Why They Succeed or Fail*, Heinemann, London.

Bessant, J. (1983) 'Management and Manufacturing Innovation: The Case of Information Technology', in Winch, G. (ed.) *Information Technology in Manufacturing Processes*, Rossendale, London.

Biondi, A. M. and Parnes S. J. (1976) *Assessing Creative Growth* (Vols. 1 and 2), The Creative Education Foundation, Buffalo, N.Y.

Brown, M. and Rickards, T. (1982) 'How to Create Creativity', *Management Today*, August.

Buijs, J. (1984) 'Stimulating Industrial Innovation – the Dutch Experience', *Creativity and Innovation Network*, Vol. 10, No 4, pp. 153–61.

Burns, T. and Stalker, G. M. (1968) *The Management of Innovation*, Tavistock, London.

Carson, J. W. (1974) 'Three Dimensional Representation of Company Business and Investigational Activities', *R. & D. Management*, Vol. 5, No. 1, pp. 35–40.

Carson, J. W. (1983) 'Repriming the Pump by Injecting New Ideas into Business', *Creativity and Innovation Network*, Vol. 9, No. 2, pp. 59–62.

Carson, J. W. (1985) 'A Strategic Model for the Facilitation of Innovation', *Creativity and Innovation Network*, Vol. 11, No. 2, pp. 72–4.

Carson, J. W. and Rickards, T. (1979) *Industrial New Product Development*, Gower, Aldershot.

Carson, J. W. and Rickards T. (1983) 'Scimitar: A Five Year Review', *Creativity and Innovation Network*, Vol. 9, No. 3, pp. 108–10.

De Bono, E. (1971) *Lateral Thinking for Management*, McGraw-Hill, Maidenhead.

Downs, G. W. R., Jun. and Mohr, L. B. (1976) 'Conceptional Issues in the Study of Innovation', *Administrative Science Quarterly*, Vol. 21, pp. 700–14.

French, W. L. and Bell, W. H. (1978) *Organization Development* (2nd edition), Prentice-Hall, Englewood Cliffs, N.J.

Freeman, C. (1982) *The Economics of Industrial Innovation* (2nd edition), Frances Pinter, London.

Gordon, W. J. J. (1961) *Synectics: The Development of Creative Capacity*, Harper & Row, New York.

Hipple, E. Von (1978) 'A Consumer-Active Paradigm for Industrial Product Idea Generation', *Research Policy*, Vol. 7, No. 2, pp. 240–66.

Janis, I. L. (1971) 'Group Think', *Psychology Today*, November.

Kirton, M. J. (1976) 'Adaptors and Innovators: A Description and Measure', *Journal of Applied Psychology*, Vol. 61, pp. 622–9.

Koestler, A. (1967) *The Act of Creation*, Hutchinson, London.

Kolb, D. A. (1984) *Experiential Learning*, Prentice-Hall, Englewood Cliffs, N.J.

Kolb, D. A., Rubin, I. M. and McIntyre, J. M. (1979) *Organizational Psychology*, Prentice-Hall, Englewood Cliffs, N.J.

Maslow, A. H. (1954) *Motivation and Personality*, Harper & Row, New York.

Mintzberg, H. (1976) 'Planning on the Left and Managing on the Right', *Harvard Business Review*, July/August.

Nelson, R. R. and Winter, S. G. (1977) 'In

Search of Useful Theory of Innovation', *Research Policy*, Vol. 6, No. 1, pp. 36–76.

Orstein, R. E. (1975) *The Psychology of Consciousness*, Jonathan Cape, London.

Osborn, A. (1963) *Applied Imagination*, Charles Scribner & Sons, New York.

Parnes, S., Noller, R. B. and Biondi, A. M. (1977) *Guide to Creative Action*, Charles Scribner & Sons, New York.

Peters, T. J. and Waterman, R. H., Jun. (1982) *In Search of Excellence*, Harper & Row, New York.

Porter, M. E. (1980) *Competitive Strategy: Techniques for Analysing Industries and Competitors*, Free Press, New York.

Rickards, T. (1974) *Problem Solving through Creative Analysis*, Gower, Aldershot.

Rickards, T. (1980) 'Designing for Creativity: A State of the Art Review', *Design Studies*, Vol. 1, No. 5 July, pp. 262–72.

Rickards, T. (1985) *Stimulating Innovation*, Frances Pinter, London.

Rickards, T. (1988) *Creativity at Work*, Gower, Aldershot.

Rickards, T. and Bessant, J. (1980) 'The Creativity Audit: Introduction of a New Research Measure', *R. & D. Management*, Vol. 10, No. 2, pp. 67–75.

Rickards, T. and Freedman, B. L. (1979) 'A Reappraisal of Creativity Techniques', *Journal of European Industrial Training*, Vol. 3, No. 1, pp. 3–8.

Rogers, E. M. (1962) *Diffusion of Innovation*, Free Press, New York.

Scheerer, M. (1963) 'Problem Solving', *Scientific American*, Vol. 208, No. 4, pp. 118–28.

Schein, E. H. (1969) *Process Consultation*, Addison-Wesley, Reading, Mass.

Schmookler, J. (1966) *Invention and Economic Growth*, Harvard University Press.

Schon, D. A. (1965) 'Champions for Radical New Inventions', *Harvard Business Review*, March-April, pp. 77–86.

Stein, M. I. (1974) *Stimulating Creativity*, Academic Press, New York.

Steiner, G. (1965) *The Creative Organization*, Chicago University Press.

Taggart, W. and Robey, D. (1981) 'Minds and Managers', *Academy of Management Review*, Vol. 6, No. 2, pp. 187–95.

Talbot, R. J. and Rickards, T. (1984) 'Developing Creativity', in Cox, C. and Beck, J. (eds) *Management Development in Advances in Practice and Theory*, Wiley, Chichester.

Taylor, C. W. and Barron, F. (1963) *Scientific Creativity*, John Wiley, New York.

Twiss, B. (1974) *Managing Technological Innovation*, Longman, London.

Utterback, J. M. (1974) 'Innovation in Industry and the Diffusion of Technology', Science Vol. 183, pp. 620–6.

Vernon, P. (ed.) (1970) *Creativity*, Penguin, Harmondsworth.

Wolfe, T. (1979) *The Right Stuff*, Farrar, Straus & Giroux, New York.

Zwicky, F. (1948) *Morphological Methods of Analysis and Construction*, Interscience, New York.

INDEX

Index

Index